The Income Approach
to Property Valuation

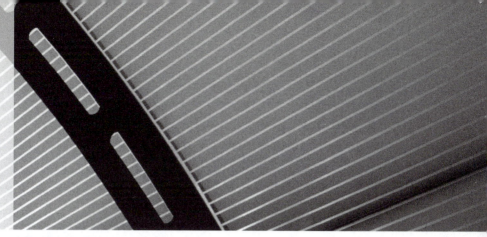

**Sixth Edition**

# The Income Approach to Property Valuation

**Andrew Baum, David Mackmin** and **Nick Nunnington**

 Books

**ELSEVIER**  AMSTERDAM BOSTON HEIDELBERG LONDON NEW YORK OXFORD
PARIS SAN DIEGO SAN FRANCISCO SINGAPORE SYDNEY TOKYO

EG Books is an imprint of Elsevier.
The Boulevard, Langford Lane, Kidlington, Oxford OX5 1GB, UK
30 Corporate Drive, Suite 400, Burlington, MA 01803, USA

Sixth edition 2011

**Notices**

The Publishers make no representation, express or implied, with regard to the accuracy of the information contained in this publication and cannot accept any legal responsibility or liability for any errors or omissions.

The material contained in this publication constitutes general guidelines only and does not represent to be advice on any particular matter. No reader or purchaser should act on the basis of material contained in this publication without first taking professional advice appropriate to their particular circumstances. The Publishers expressly disclaim any liability to any person who acts in reliance to the contents of this publication. Readers of this publication should be aware that only Acts of Parliament and Statutory Instruments have the force of law and that only courts can authoritatively interpret the law.

**British Library Cataloguing-in-Publication Data:** 332.6'324'0941-dc22

**Library of Congress Cataloging-in-Publication Data:** 2011920620

ISBN: 978-0-08-096690-8

For information on all EG Books publications
visit our website at www.elsevier.com

Printed and bound in Great Britain.

11 12 13   10 9 8 7 6 5 4 3 2 1

Working together to grow
libraries in developing countries

www.elsevier.com | www.bookaid.org | www.sabre.org

ELSEVIER    BOOK AID International    Sabre Foundation

# Contents

# Preface to Sixth edition

This edition of the book begins with an introduction and a quick start to the income approach; this is followed by a consideration of the investment arithmetic which underpins the income approach and a review of the basic principles of valuation. The application of the approach to the assessment of the market value of freehold and leasehold investments is then considered, before looking at the impact of legislation on what would otherwise be a relatively unfettered market. Finally, the book covers a number of specialist areas of valuation relating to the use of the profits version of the income approach, development properties and investment analysis. Throughout the book it is assumed that the reader has some knowledge of who buys property, why they buy it and what alternative investment opportunities there are, and also that the reader will have some knowledge of the nature of property as an investment. The reader should have some awareness of the social, economic and political factors that influence the market for and the value of property. For further consideration of these issues readers are referred to the sister publication, *Principles of Valuation*, also by EG Books (an imprint of Elsevier).

In preparing the sixth edition, we have taken note of comments and reviews submitted to our editor at Elsevier. In particular, we have reflected on the dual rate battle and have removed almost all reference to a method which we have always had concerns over. Those who miss it will find support in the Appendices for their teaching and learning. We have tried to enhance the material so as to leave readers with fewer puzzles of 'how did they do that' or simply 'why that'? We have extended some of the spreadsheet material and uses of standard software; but left much to the reader to discover for his or herself as finding Excel® solutions to valuation tasks is, in itself, a powerful learning tool. Leasehold Enfranchisement, which kept growing, has gone to where it belongs in the main statutory valuation text books. One weakness has been our consideration of the Profits Method. This has been addressed in this edition with a new chapter by Howard Day of Howard Day Associates Ltd, Chartered Surveyors providing property advice to the leisure sector.

We have sought to use terminology which is consistent with the Royal Institution of Chartered Surveyors' (RICS) *Valuation Standards* (the Red Book) and which reflects the international use of the income approach. A particular change is the adoption, for the most part, of the internationally recognised Present Value of £1 per annum (PV£1 pa) in place of the UK valuers' Years' Purchase. Whilst valuers in the United Kingdom do use the 12th edition of *Parry's* Valuation and Investment Tables, most students now use calculators, Excel® and valuation software; to conform to this we refer to the financial functions rather than to tables.

As in previous editions, all examples are for illustration only and are not intended to be a reflection of current market rents and yields.

The art or science of valuation has developed since the first edition and our purpose now is to provide a comprehensive review of the income approach in as simple a style as possible, exploring the wide range of opinions and views that have been expressed about the how, why and when of the methodology. For practitioners, we would ask that if you see an approach which appears to be a new technique to you then please keep an open mind; try it and see, but remember market value is your opinion, not just the mathematical result of the income approach you chose to use.

Andrew Baum
David Mackmin
Nick Nunnington.
Reading, Sheffield, Abu Dhabi
2011

# Acknowledgements

Malcolm Martin BSc, FRICS, FNAEA provided much needed assistance in the fifth edition on leasehold enfranchisement which has now gone, but we have made use of the revisions he made to the section on residential valuation. However, if we now have it wrong please blame us, not Malcolm.

Andrew and David still recall the enthusiasm for their ideas in 1979 from Peter Byrne and David Jenkins and will continue to say thank you.

A major change in this edition has been the recognition of the growth in the leisure sector of the market and the coming of age of the profits approach. We have responded with a much enlarged chapter on this method which has been written for us by Howard Day BSc (Hons.) FRICS, MAE, MCIArb, FAVLP of Howard Day Associates Ltd, Liberty House, London W1B 5TR, Chartered Surveyors providing property advice to the leisure sector.

Extracts from the RICS Valuation Standards (2011 edition), the Red Book, are RICS copyright and are reprinted here with their permission. Readers are advised that the 7th edition of the Red Book is effective from 2nd May 2011.

Argus Developer software has been used in Chapter 10 and Appendix C to illustrate various aspects of development appraisal.

# Chapter  1

## Introduction and Quick Start to the Income Approach

## Introduction

Cairncross (1982), in his *Introduction to Economics*, expresses his view that 'economics is really not so much about money as about some things which are implied in the use of money. Three of these - exchange, scarcity and choice - are of special importance'. Legal interests in land and buildings, which for our purposes will be known as property, are exchanged for money and are scarce resources. Those individuals fortunate to have surplus money have to make a choice between its alternative uses. If they choose to buy property they will have rejected the purchase of many other goods, services, alternative investments such as stocks and shares, or of simply leaving it in a savings account or doing nothing. Having chosen to use their surplus money to purchase property, they will then have to make a choice between different properties. Individuals investing in pension schemes, endowment or with profits life assurance policies are entrusting their money to others to make similar choices on their behalf.

Valuation is the vocational discipline within economics that attempts to aid that choice in terms of the value of property and the returns available from property. Value in this context can mean the market value in exchange for property rights, as well as value to a particular person or institution with known objectives, currently referred to as an appraisal, or assessment of worth, or investment value. Valuation was defined in the Royal Institution of Chartered Surveyors (RICS) *Valuation Standards* (The Red Book) (2010) as:

> A valuer's opinion of the value of a specified interest or interests in a *property*, at the *date of valuation*, given *in writing*. Unless limitations are agreed in the *terms of engagement* this will be provided after an inspection, and any further investigations and enquiries that are appropriate, having regard to the nature of the *property* and the purpose of the *valuation*.

A valuer in the context of this book would be a member of the RICS or Institute of Revenues Rating and Valuation (IRRV) who meets the qualification requirements specified in Valuation Standard (VS) 1.4 and has the knowledge and skills specified in VS 1.6, and who acts with independence, integrity and objectivity as specified in VS 1.7.

The RICS Valuation Standards adopt the International Valuation Standards (IVS) definition of market value, namely:

The estimated amount for which an asset or liability should exchange on the *valuation date* between a willing buyer and a willing seller in an arm's–length transaction after proper marketing wherein the parties had each acted knowledgeably, prudently and without compulsion. (RICS VS 3.2).

It also adopts investment value as:

The value of an asset to the owner or a prospective owner. (May also be known as worth.) (RICS VS Glossary)

At the time of preparing this edition, the International Valuation Standards Council (IVSC) has a consultation draft out for a new edition of the IVS and the RICS 7th edition of The Red Book, which is compliant with IVS, is scheduled for publication in April 2011 with an effective date of 2nd May 2011. Readers are advised to refer to the new edition after publication as it is compulsory reading for valuers and there may be changes which, as yet, we have not been able to incorporate. The 7th edition is different; for example the word property in the definition of MV has been replaced with asset as MV should be defined consistently for all asset classes not just property.

Property is purchased for use and occupation or as an investment. In both cases the purchaser measures the expected returns or benefits to be received from the property against the cost outlay. The valuer's task is to express these benefits in monetary terms and to interpret the relationship between costs and benefits as a rate of return, thus allowing the investor to make a choice between alternative investment opportunities.

Since 1945, the property and construction industries have grown in importance; investing in property has been indiscriminately considered to be a 'safe' investment. The position post the banking crisis in 2007 aptly illustrates how dangerous it can be to make such an assumption. The growth in pension schemes, life funds, property unit trusts, as well as direct investment by individuals, has completed the transition of the property market into a multi-billion pound industry. As a result, there has been a growth in demand for property to be valued in order to establish market value and for it to be re-valued for portfolio and asset management purposes.

Property as an investment is different to other forms of investment. The most obvious difference is its fixed location geographically, hence the importance of the quality of that location for the land's current or alternative uses as determined by its general and special accessibility, and its interrelationship with other competing and complementary buildings, locations and land uses. Once developed, the quality of the investment is influenced by the quality of the permitted planning use and the quality of the physical improvements (buildings) on the site. In addition and essential to the assessment of exchange value is the quality of the legal title. Is it freehold or leasehold? The owner of a freehold title effectively owns all the land described in the title deeds in perpetuity, including everything below it to the centre of the earth and everything above. Freehold rights may be restricted by covenants in the title and/or by the rights of others, such as rights of way. A leaseholder's rights are limited in time (the length of the lease) and by the terms and conditions (covenants) agreed between landlord and tenant and written into the lease, or implied or imposed by law or statute. The market value of a tenanted freehold property will also be affected by the quality of the tenant in terms of their covenant strength and the quality of the lease in terms of the appropriateness of the lease conditions to that type of property used for that purpose in that location.

To be competent, the valuer must be aware of all the factors and forces that make a market and which are interpreted by buyers, sellers and market makers in their assessment of market price. In an active market where many similar properties with similar characteristics and qualities are being exchanged, a valuer will, with experience, be able to measure exchange value by comparing that which is to be valued with that which has just been sold. This direct or comparative method of valuation is used extensively for the valuation of vacant possession, freehold, residential property; for the valuation of frequently sold and easily compared commercial and industrial property and to assess the market rent of all property. Differences in age, condition, accommodation and location can all, within reason, be reflected by the valuer in the assessment of value. Differences in size can be overcome by adopting a unit of comparison such as price per hectare, price per square metre or rent per square metre.

The more problematic properties are those for which there is no ready market, those which display special or unique characteristics, those which do not fully utilise the potential of their location and are therefore ripe for development, redevelopment or refurbishment, those that are tenanted and are sold as investments at prices reflecting their income generating potential; and leasehold properties.

For each of these broad categories of property, valuers have developed methods of valuation that they feel most accurately reflect the market's behavioural attitude and which may therefore be considered to be rational methods.

In the case of special properties such as oil refineries, glassworks, hospitals and schools, the usual valuation method is the cost or contractor's method. It is the valuer's method of last resort, and is based on the supposition that no purchaser would pay more for an existing property than the sum of the cost of buying a similar site and constructing a similar building with similar utility written down to reflect the physical, functional, environmental and locational obsolescence of the actual building. This approach is sometimes referred to as a Depreciated Replacement Cost approach, or DRC for short.

Properties with latent development value are valued using the residual or development (developer's) method (see Chapter 10). The logic here is that the value of the property (site) in its current state must equal the value of the property in its developed or redeveloped state, less all the costs of development including profit but excluding the land. In those cases where the residual sum exceeds the value in its current use, the property is considered ripe for development or redevelopment and, in theory, will be released for that higher and better use.

All property that is income producing or is capable of producing an income in the form of rent, and for which there is both an active tenant market and an active investment market, will be valued by the market's indirect method of comparison. This is known as the investment method of valuation or the income approach to property valuation and is the principal method considered in this book.

The income approach and the income-based residual approach warrant special attention if only because they are the valuer's main tools for valuing the most complex and highly priced investment properties.

Real estate (property) is an investment. There are three main investment asset classes – stocks and corporate bonds, equity shares, and property.

An investment can be described as an exchange of capital today for future benefits, generally in the form of income (dividends, rent, etc.) and sometimes in the form of capital. The investment income from property is the net rent paid by tenants. The market price of an investment is determined in the market by the competitive bids of buyers for the available supply at a given point in time under market conditions prevailing at that time. Short supply and high demand that is scarcity will lead to price (value) increases, whilst low demand and high supply will lead to falls in prices and values.

The definition of market value requires the valuer to express an opinion of the market price that the valuer believes would have been achieved if that property had been sold at that time under the market conditions at that time.

The unique characteristics of property make property investment valuation more complex an art and science than that exercised by brokers and market makers in the market for stocks and shares. In the stock market, sales volume generally allows for price (value) comparison to be made minute by minute. As stocks, shares and property are the main investments available, there is bound to be some similarity between the pricing (valuation) methods used in the various markets and some relationship between the investment opportunities offered by each. A basic market measure is the investment yield or rate of return. The assessment of the rate of return allows or permits comparison to be made between investments in each market and between different investments in different markets. There is a complex interrelationship of yields and patterns of yields within the whole investment market. In turn these yields reflect market perceptions of risk and become a key to pricing and valuation methods. Understanding market relationships and methods can only follow from an understanding of investment arithmetic.

## The income approach – a quick start

The income approach or investment method of valuation is an internationally recognised method of assessing market value of property. Buyers of property require an acceptable return or yield on their invested money. The yield must be sufficient to compensate for the risks of exchanging money now with today's purchasing power for future income with a future uncertain purchasing power.

The initial yield is a simple measure of the income/capital relationship.

$$\frac{Income(I)}{Purchase\ Price(P)} \times 100 = Yield\ (r\%)$$

If a buyer pays £2,000,000 for a property producing a net rent of £100,000 which is considered to be the market rent, then this sale price can be analysed to find the yield on money invested.

$$\frac{£100,000}{£2,000,000} \times 100 = 5\%$$

The market yield reflects all the risks perceived by the players in the market at the time of the purchase. These market yields provide the valuer with a key measure of an investment and a key tool for the income approach.

The experienced valuer with market knowledge of the risks associated with investing in property in general and with those of a specific property can arrive at an opinion, by comparison, of the yield that buyers would require from a given property in order for a purchase to occur. Risks to be considered will relate to: the legal title; the physical construction and condition of the property; the location of the property; the use of the property; the quality of the occupying tenant, i.e. their covenant strength; the length of the lease; the lease covenants and many other factors.

Given that $\frac{I}{P} \times 100 = r\%$ *then* $\frac{I}{\left(\frac{r}{100}\right)} = P$ and so the value of a similar property let at its market rent of £59,500 could be calculated as

$$\frac{£59,500}{\left(\frac{5}{100}\right)} = \frac{£59,500}{0.05} = £1,190,000$$

In the income approach this is called capitalising the income or the capitalisation approach. The value of £1 of income will vary with the yield and the yield will vary with the perceived investment risks. The importance of the yield can be seen in the following table; here a fall of 1% from 12% to 11% is a 9.12% increase in the present value of £1 pa in perpetuity, but a 1% fall from 5% to 4% increases the present value of £1 in perpetuity by 25%.

| Income | Yield r% (100/r)*† | Present Value of £1 pa in perpetuity (PV£1 pa in perp). | Change in value per 1% fall in yield |
|---|---|---|---|
| £1 | 4% (25.00) | £25 | 25% |
| £1 | 5% (20.00) | £20 | 19.98% |
| £1 | 6% (16.67) | £16.67 | 16.66% |
| £1 | 7% (14.29) | £14.29 | 14.32% |
| £1 | 8% (12.50) | £12.50 | 12.51% |
| £1 | 9% (11.11) | £11.11 | 11.10% |
| £1 | 10% (10.00) | £10.00 | 10.01% |
| £1 | 11% (9.09) | £9.09 | 9.12% |
| £1 | 12% (8.33)* | £8.33 | – |

*These figures are rounded to two decimal places but for valuation need to be calculated to four decimal places.
†UK valuers use the term Years' Purchase (YP) for the product of (100/r) or (1/$i$) which is then used as a multiplier to turn income in perpetuity into its present value equivalent.

Capitalisation in its simplest form where income can be assumed to be perpetual is simply Income ÷ i% where i% is (r ÷ 100). So at 4% it is £1 ÷ 0.04 = £25 or £1 × (1 ÷ 0.04) = £1 × 25YP = £25. Capitalisation is a market-based method of valuation which requires knowledge of market rents derived from analysis of lettings of property and of market yields derived from analysis of market sales.

Valuation mathematics is not always this simple. In many situations the rent being paid is not the market rent and account must be taken of the rent actually being paid and for how long, plus the reversion to market rent at the next rent review or lease renewal.

Capitalisation is a short-cut discounted cash flow (DCF) or present value calculation. Where the income changes over time, the valuer needs to reflect the time value of money through the use of present values.

Investors can save money or invest money. Money saved earns interest, and interest compounds over time if not withdrawn. The general equation for compound interest is $(1 + i)^n$ where i is the rate of interest expressed as a decimal, i.e. in respect of £1 not as a percentage (£100) and n is the number of interest earning periods. So:

$$£\,1000 \times (1 + 0.10)^{10} = £\,2{,}593.74$$

[Note that]
$$(1.10)^{10} = (1.10) \times (1.10) \times (1.10) \times (1.10) \times (1.10)$$
$$\times (1.10) \times (1.10) \times (1.10) \times (1.10) \times (1.10)$$

A thousand pounds saved today and left to earn interest at 10% a year will compound over 10 years to £2,593.74. In which case

£2,593.74 due to be received in 10 years time will have a present value today of:

$$\pounds\,2{,}593.74 \times \frac{1}{(1+i)^n} = \pounds2{,}593.74$$
$$\times \frac{1}{(1+0.10)^{10}} = \pounds\,2{,}593.74 \times 0.3855433 = \pounds\,1{,}000$$

The value of £1 due in 10 years time has a PV today at 10% of £0.3855433.

Value a property let at £80,000 a year until a rent review in three years to market rent which today is £100,000. Market yields are 8%. This calculation can be performed easily with a standard calculator or preferably with a financial calculator such as an HP 10bII.

The cash flow is:

| End year | Income (cash flow) |
|---|---|
| 1 | £80,000 |
| 2 | £80,000 |
| 3 | £80,000 |
| Year 3 – perpetuity | £1,250,000 (£100,000 ÷ 0.08) |

Therefore, the present value at 8% is:

| | | |
|---|---|---|
| 100,000 ÷0.08 | £1,250,000 | |
| Plus £80,000 end year 3 rent | £1,330,000 ÷ 1.08 | £1,231,481.48 |
| Plus £80,000 end year 2 rent | £1,311,481.48 ÷ 1.08 | £1,214,334.71 |
| Plus £80,000 end year 1 rent | £1,294,334.71 ÷ 1.08 | **£1,198,458.06** |

OR, using a Hewlett-Packard HP10bII calculator (here $f$ indicates the need to use the function key):

| Keyboard | Screen display | Description |
|---|---|---|
| $f$ C All | 0.00 | Clears all registers |
| 1 $f$ P/YR | 1.00 | Stores payment periods per year; here it is a single payment |
| 0($\pm$)(*CFj*) | −0.00 | Creates a day one zero receipt to achieve in arrears income payments |
| 80,000 | CFj | End year 1 rent |
| 80,000 | CFj | End year 2 rent |
| 1,330,000 | CFj | End year 3 rent plus 100,000 in perpetuity at 8% |
| 8 | 1/YR | 8% nominal yield to be used |

*(Continued)*

| Keyboard | Screen display | Description |
|---|---|---|
| ƒ NPV | 1,198,458.06 | Is therefore the present value of £80,000 for three years, plus £100,000 in perpetuity beginning after three years all discounted at 8%. |

As the income is constant for a number of years, there is no need to enter as single cash flows, a quicker solution is possible.

The standard textbook valuation layout would appear as:

| Term | Calculation | Present Value |
|---|---|---|
| Current rent | £80,000 | |
| PV of £1 pa (YP)at 8% for 3 years | 2.5771 | £206,167.76 |
| **Reversion** | | |
| Market rent | £100,000 | |
| PV of £1 pa in perp at 8% (YP perp) | 12.50 | |
| | £1,250,000 | |
| PV of £1 in 3 years at 8% | 0.7938322 | £992,290.25 |
| Sum term and reversion | | £1,198,458.01 |

Note that the PV of £1 pa is the sum of the PV of £1 for 1 year plus PV £1 for 2 years plus PV £1 for 3 years. The equation is $\dfrac{1 - \dfrac{1}{(1+i)^n}}{i}$.

Kel Computing Ltd (www.kel.co.uk) have for several years provided a free student download of their Investment Valuer. This can be accessed and downloaded from www.keldownloads.co.uk/downloads/student/KELInvestmentValuer - Student Version 2010-11. EXE or http://tinyurl.com/studentKel.

If you are using the Kel download then look at the content of the various tabs first; then load the valuation data; then click on report and valuation. Kel will become much clearer after Chapter 5. If you are on a full-time course of study at university or college then you may have access to Argus software or one of the other valuation packages used by the profession. However, at this stage, Kel Investment Valuer for students is possibly an easier starting tool. Downloading is probably only possible on a home computer or laptop as most large organisations only allow authorised staff to download from the internet.

To benefit from this quick start you need to complete the following questions using one or other of the methods set out above. Your approach will depend upon your calculator and its functions. If your answers agree with those shown then you have discovered how to find the present value of a given cash flow at a given yield or discount rate. Now all that is needed is an understanding of where the cash flow comes from in respect of a given property, and how to deduce the correct yield. In passing, two approaches to the income approach will be explored – income capitalisation and DCF.

### Questions

A. Valuation problems where the current rent being paid is the market rent and is a net rent.

  1. A freehold interest in a shop where the unit has been let at the market rent of £75,000. Market yields would support a valuation at 6%. (Answer £1,250,000.)
  2. A freehold interest in a business park consisting of four identical units. Each is let at the market rent of £45,000. Market evidence would support a valuation on a 7.5% basis. (Answer £2,400,000.)
  3. A freehold interest in an office building let at the market rent of £1,555,000. Market yields for this quality of office investment are 7%. (Answer £22,214,285.)

B. Valuation problems where the property is not let at its market rent but there is a rent review in 'n' years' time to market rent.

  4. A freehold interest in a shop unit let at £65,000 a year with a rent review in two years time to a market rent of £75,000. Comparable market yields are 6%. (Answer £1,231,666.)
  5. A freehold interest in a business park with four identical units. Each is let at a rent of £39,500. There are rent reviews in three years time on each unit. The market rent is £45,000. Comparable market yields are 7.5%. (Answer £2,342,767.)
  6. A freehold interest in an office building let at £1,332,000 a year for the next five years. This was an incentive rent to secure an early and quick letting of the building. The market rent is £1,555,000 and the rent review is in five years time. Comparable market yields are 7%. (Answer £21,299,950.)

The following chapters explore the processes and applications of the income approach in more detail, including the various tools or techniques preferred by different valuers.

# Chapter 2

## Financial Mathematics for Valuers

## Introduction

This book explores the process of property valuation, with particular reference to property which is bought or sold as an investment. In order to be able to value an investment property, a valuer must understand how the benefits to be enjoyed from the ownership of a freehold or leasehold interest in property can be expressed in terms of present value (PV).

To do this, a valuer must have a working knowledge of the mathematics of finance and the theory of compounding and discounting as it relates to savings and investments.

This chapter explores the mathematics behind the six investment functions of £1 and other derived functions, and illustrates their use in the practice of property valuation.

## The six functions of £1

The six basic functions of financial mathematics are set out below; throughout $i$ is the rate of interest expressed as a decimal, e.g. 5% = 5/100 = 0.05, this must always be the correct rate per interest earning period (see nominal and effective rates later in this chapter); $n$ is the number of interest earning periods.

Percentages are used frequently by valuers. As a reminder, one per cent (1%) is one hundredth part of the whole, or $1/100$th. If a gift of £100 is to be divided equally between 10 people then it has to be divided into 10 parts; each person will receive $1/10$th or 10% [$1/10$th being $(1/10 \times 100) = 10\%$]. So $1/10$th or 10% of £100 is [(£100/100) × 10] which is £10.

**The Amount of £1 (A)** another name for compound interest; that is, the future worth of £1, if invested today allowing for compound interest at a given rate.

$$A = (1 + i)^n$$

**The Amount of £1 per annum (A £1 pa)** that is, the future worth of £1 invested at the end of each year accruing compound interest at a given rate.

$$A\,£1\mathrm{pa} = \frac{(1 + i)^n - 1}{i}$$

**The Annual Sinking Fund (ASF)** that is, the fraction of £1 which must be invested at regular intervals to produce £1 at a given point in the future with compound interest accruing at a given rate.

$$ASF = \frac{i}{(1+i)^n - 1}$$

**The PV of £1 (PV £1)** that is, the PV of £1 to be received in the future, discounted over a given period at a given rate.

$$PV = \frac{1}{(1+i)^n} = \frac{1}{A}$$

**The PV of £1 per annum (PV £1 pa)** that is, the PV of a series of payments of £1 due annually for a given period of time, discounted at a given rate. This function is also known by valuers as the Year's Purchase (YP) single rate.

$$PV\ £1pa = \frac{1 - \dfrac{1}{(1+i)^n}}{i} = \frac{1 - PV}{i}$$

**The Annuity £1 Will Purchase (AWP)** that is, the amount of money that will be paid back at the end of each year in return for £1 invested today for a given number of years at a given rate.

$$A£1pa = i + ASF$$

There are several published valuation tables such as *Parry's Valuation and Investment Tables*, 12th edition (2002), published by EG Books (Elsevier), which show the value of these and other functions for a range of interest rates and years. Some of these tables compute income as being received quarterly in advance, as with many actual investments. However, in order to simplify this introductory chapter it is assumed initially that income is received or invested at the beginning or end of each year in one instalment.

Whilst reference is made throughout to pounds Sterling (GBP £), the principles apply to any currency; and the Euro or US Dollar or any other currency can be substituted.

## The amount of £1(A)

The amount of £1 is simply another term for compound interest. Consider the building society passbook, set out below, where the interest rate for deposits is at 10% per annum.

| Date | Description | Deposits | Withdrawals | Interest | Balance |
|------|-------------|----------|-------------|----------|---------|
| 01/01/12 | Cash | £100 | | | |
| 31/12/12 | Interest | | | £10.00 | £110.00 |
| 31/12/13 | Interest | | | £11.00 | £121.00 |

| Date | Description | Deposits | Withdrawals | Interest | Balance |
|---|---|---|---|---|---|
| 31/12/14 | Interest | | | £12.10 | £133.10 |
| 31/12/15 | Interest | | | £13.31 | £146.41 |

This table shows that the interest accumulates on both the original £100 invested and on the interest added to it.

The formula to express this states that if £1 is invested for one year at $i$ interest, then at the end of one year it will have accumulated to $(1 + i)$. At 5% this will read as $(1 + 0.05)$ and the initial £1 with interest added becomes £1.05.

At the end of the second year $(1 + i)$ will have earned interest at $i\%$, so at the end of the second year it will have accumulated to:

$$(1 + i) + i(1 + i) \text{ which can be expressed as} (1 + i)^2$$

At the end of $n$ years the accumulated sum will be;

$$(1 + i)^n$$

Remember that interest is expressed as a decimal ($i$) or ($r$), e.g. 10% = 10 divided by 100 = 0.10 and that whilst $n$ is normally the number of years of interest accumulation, the formula or equation is a general one for compound interest. So, if interest is added monthly, then $n$ for one year becomes 12 and $i$ will be the monthly rate of interest.

---

**Example 2.1**

Calculate the amount of £1 after four years at 10%.

$A = (1 + i)^n \qquad i = 0.10; n = 4$
$A = (1.1)^4$
$A = (1.1) \times (1.1) \times (1.1) \times (1.1)$
$A = 1.4641$

The calculation shows that £1 will accumulate to £1.4641 (£1.47 to the nearest pound and pence) after four years at 10% compound interest rate. Notice that if the figure produced by the formula in example 1.1 (1.4641) is multiplied by 100, the figure is the same as that shown in the building society passbook.

The formula for the amount of £1 is therefore;

$$A = ([1 + i])^n$$

## The amount of £1 per annum (A £1 pa)

This function is used to calculate the future value, with compound interest, of a series of payments made at regular intervals, normally each year. The Latin 'per annum' or, as shortened, pa, is the expression normally used for annual receipts or payments. Many investments follow this pattern. Consider for example another building society passbook set out below where this time the investor deposits £100 at the end of each year 01,02,03,04,05 and interest is added at 10% per annum.

| Date | Description | Deposits | Withdrawals | Interest | Balance |
|------|-------------|----------|-------------|----------|---------|
| 01 | Cash | £100 | | | |
| 02 | Interest | | | £10.00 | £110.00 |
| 02 | Cash | £100 | | | £210.00 |
| 03 | Interest | | | £21.00 | £231.00 |
| 03 | Cash | £100 | | | £331.00 |
| 04 | Interest | | | £33.10 | £364.10 |
| 04 | Cash | £100 | | | £464.10 |
| 05 | Interest | | | £46.41 | £510.51 |
| 05 | Cash | £100 | | | £610.51 |

The amount of £1 pa deals with this type of investment pattern; it indicates the amount to which £1, invested at the end of each year, will accumulate at $i$ interest after $n$ years.

The table is simply a summation of a series of amounts of £1. If each £1 is invested at the end of the year, the $n$th £1 will be invested at the end of the $n$th year and will thus earn no interest. See last cash entry above for 05.

Each preceding £1 will earn interest for an increasing number of years:

- The $(n-1)$ £1 will have accumulated for 1 year, and will be worth $(1+i)$.
- The $(n-2)$ £1 will have accumulated for two years, and will be worth $(1+i)^2$.
- The first £1 invested at the end of the first year will be worth $(1+i)^{n-1}$.

This series of calculations when added together is expressed as:

$$1+(1+i)+(1+i)^2 \ldots (1+i)^{n-1}$$

This is a geometric progression and when summed it can be expressed as:

$$\frac{(1+i)^n - 1}{i}$$

This is the formula for the amount of £1 pa.

---

**Example 2.2**

Calculate the amount of £1 per annum for five years at 10%.

$$A\,£\,1\text{pa} = \frac{(1+i)^n - 1}{i} \qquad i=0.10\,;\,n=5$$

$$A\,£\,1\text{pa} = \frac{(1.10)^5 - 1}{i} = \frac{1.61051 - 1}{0.10}$$

$$A\,£\,1\text{pa} = \frac{0.61051}{0.10} = 6.1051$$

Multiplying this figure by 100 to calculate the sum that £100 invested at the end of each year will accumulate to after five years produces the same figure as in the building society passbook.

**Example 2.3**

If Mr A invests £60 in a building society at the end of each year, and at the end of 20 years has £4,323, at what rate of interest has this sum accumulated?

| | |
|---|---|
| Annual sum invested | £60 |
| A £1 pa for 20 years at i % | x |
| Capital Value (CV) of Investment at the end of 20 years is | £4,323 |

Rephrasing as a simple equation:
£60x = £4,323, i.e. £60 multiplied by an unknown number x will produce a sum of £4,323 and so:

$$x = \frac{£4{,}323}{£60} = 72.05$$

Substituting in the formula for the A £1 pa and solving to find *i* or if the reader prefers by checking in *Parry's* Amount of £1 pa table, it will be seen that 72.05 is the value for 20 years at 12% which is the rate of compound interest at which this regular investment has accumulated.

# Annual sinking fund (ASF)

Property investors may need to make provision from today for some future expenditure. This may be for some form of maintenance, for example, a building may require a new roof or service roads may require resurfacing.

Such obligations may be passed on to the tenant in the form of service charges, payable in addition to rent, which include the provision of a fund to be used for major works in the future. However, in some cases, the landlord may be responsible for major repairs which cannot be recovered from the tenant. Such obligations should be reflected in the purchase price of the investment. Such expenditure can be budgeted for by investing a lump sum today or by investing a regular sum from today, both of which will grow due to compound interest. A property owner may need to know how much should be invested now or on an annual basis.

The first step is to estimate the probable amount needed at an expected future date. The investor could then either invest a lump sum immediately which, with the accumulation of interest, will meet the estimated outlay when it arises. This can be calculated using the PV of £1 (see later). However, it may be more effective to set aside part of the income received from the investment (i.e. rent) regularly in an account known as a sinking fund, which is planned to accumulate to the required sum by the date in the future when the expenditure is required.

This is similar to the amount of £1 pa described above. Example 2.4 demonstrates how the annual sinking fund (ASF) and A £1 pa can be used to calculate the sum required to be set aside in a sinking fund to meet a known future expense.

### Example 2.4

An investor is considering the purchase of a small shop in which the window frames have begun to rot. It is estimated that in four years time they will require complete replacement at a cost of £1,850. The shop produces a net income of £7,500 pa.

How much of this income should be set aside each year to meet the expense, assuming the money is invested with a guaranteed fixed return of 3% per year?

The ASF is the reciprocal of A £1 pa. The use of ASF enables this sum to be calculated easily.

Calculate the ASF to accumulate to £1 after 4 years at 3% given that the ASF is:

$$ASF = \frac{i}{(1+i)^n - 1}$$

$$ASF = \frac{i}{(1+i)^n - 1} = \frac{0.03}{(1.03)^4 - 1} = \frac{0.03}{1.1255088 - 1} = 0.0239027$$

| | |
|---|---|
| Sum required | £1,850 |
| X by ASF to replace £1 in 4 years at 3% | 0.239027 |
| ASF | £442.20 |

It can be seen that this calculation performs the function of the amount of £1 pa in reverse. The annual sum was found by multiplying the sum required by the ASF to replace £1 at the end of four years with compound interest at 3%. Using the A £1pa, the sum of £1,850 could be divided by the A £1pa for four years at 3%, thus £1,850/4.1836 = £442.20. Therefore, as mentioned above, the ASF is the reciprocal of the amount of £1 pa:

$$ASF = \frac{i}{(1+i)^n - 1}$$

## The present value of £1 (PV £1)

The first three functions of £1 have shown how any sum invested today will be worth more at some future date due to the accumulation of compound interest. This means that £1 receivable in the future cannot be worth the same as £1 at the present time. What it is worth will be the sum that could be invested now to accumulate to £1 at a given future date at a given rate of interest. This sum will obviously depend upon the length of time over which it is invested and the rate of interest it attracts.

If £1 were invested now at a rate of interest $i$ for $n$ years, then at the end of the period it would be worth $(1 + i)^n$.

If £x was to be invested now at $i\%$ for $n$ years and it accumulates to £1:

$$If\ £x \times (1+i)^n = £1\ then\ £x = £1 \times \frac{1}{(1+i)^n}$$

This is the formula for the PV of £1 and it is the reciprocal of the Amount of £1.

$$PV = \frac{1}{(1+i)^n} = \frac{1}{A}$$

Proof:

- PV £1 in 7 years at 10% = 0.51316 $[1/(1.10)^7]$.
- A £1 in 7 years at 10% = 1.9487 $[(1.10)^7]$.
- And 0.51316 × 1.9487 = 1.00.

### Example 2.5

If X requires a return of 10% pa, how much would you advise X to pay for the right to receive £200 in five years time?

$$PV = \frac{1}{(1+i)^n} = \frac{1}{(1+.10)^5} = \frac{1}{1.16105} = 0.6209 \times £\,200 = £\,124.18$$

This means that if £124.18 is invested now and earns interest at 10% each year then it would accumulate with compound interest to £200 in five years time. £124.18 is the present value of £200 due to be received in five years time at 10%.

## The PV of £1 pa (PV £1 pa or Years' Purchase)

The amount of £1 pa was seen to be the summation of a series of amounts of £1. Similarly, the PV of £1 pa is the summation of a series of PVs of £1. It is the PV of the right to receive £1 at the end of each year for $n$ years at $i\%$.

The PV of £1 receivable in one year is:

$$\frac{1}{(1+i)}$$

in two years it is:

$$\frac{1}{(1+i)^2}$$

and so the series reads:

$$\frac{1}{(1+i)} + \frac{1}{(1+i)^2} + \frac{1}{(1+i)^3} \cdots \frac{1}{(1+i)^n}$$

This is a further geometric progression which when summated can be expressed as:

$$\frac{1 - \dfrac{1}{(1+i)^n}}{i}$$

This is the formula for the PV of £1 pa, and if:

$$\frac{1}{(1+i)^n} = PV$$

then the PV £1 pa can be simplified to:

$$\frac{1 - PV}{i}$$

---

**Example 2.6**

Calculate the PV of £1 pa at 5% for 20 years given that the PV of £1 in 20 years at 5% is 0.3769

$$PV£1\ pa = \frac{1 - PV}{i} = \frac{1 - 0.3769}{0}.05 = 0.\frac{6231}{0}.05 = 12.462$$

---

**Example 2.7**

How much should A pay for the right to receive an income of £675 for 64 years if A requires a 12% return on the investment?

| Income per year | £675 |
|---|---|
| X PV £1 pa for 64 years at 12% | 8.3274 |
| PV (or value today) | £5,621 |

The PV of £1 pa is referred to by UK property valuers as the 'Years' Purchase' (YP). The *Oxford English Dictionary* gives a date of 1584 for the first use of this phrase '*at so many years' purchase*', which was used in stating the price of land in relation to the annual rent in perpetuity. This term is sometimes confusing as it does not relate to the other investment terms. However, the terms are interchangeable and both are used by valuers, but internationally the more acceptable term is PV £1 pa.

Obviously, the PV £1 pa will increase each year to reflect the additional receipt of £1. However, each additional receipt is discounted for one more year and will be worth less following the PV rule established above. The PV £1 pa in fact approaches a maximum value at infinity. However, as the example below shows, the increase in PV £1 pa becomes very small after 60 years and is customarily assumed for the purpose of property valuation to reach its maximum value after 100 years. In valuation terminology this is referred to as 'perpetuity'.

**Example 2.8**

In the formula $\dfrac{1-PV}{i}$ what happens to PV as the time period increases?

What effect does this have on the value of the PV £1 pa?

From Table 2.1, two facts are clear, the PV decreases over time and the PV £1 pa increases over time. In addition it can be seen that the PV £1 pa is the accumulation of the PVs.

As $n$ approaches perpetuity, the PV tends towards 0; the PV of £1 to be received such a long time in the future is reduced to virtually nothing.

**Table 2.1**

| Years | PV at 10% | PV £1 pa at 10% |
|---|---|---|
| 1 | 0.90909 | 0.90909 |
| 2 | 0.82645 | 1.736 |
| 3 | 0.75131 | 2.487 |
| 4 | 0.68301 | 3.170 |
| 5 | 0.62092 | 3.791 |
| 6 | 0.56447 | 4.355 |
| 7 | 0.51316 | 4.868 |
| 8 | 0.46651 | 5.335 |
| 9 | 0.42410 | 5.759 |
| 10 | 0.38554 | 6.145 |
| | | |
| 90 | 0.0001882 | 9.998 |
| 91 | 0.0001711 | 9.998 |
| 92 | 0.0001556 | 9.998 |
| 93 | 0.0001414 | 9.999 |
| 94 | 0.0001286 | 9.999 |
| 95 | 0.0001169 | 9.999 |
| 96 | 0.0001062 | 9.999 |
| 97 | 0.0000966 | 9.999 |
| 98 | 0.0000878 | 9.999 |
| 99 | 0.0000798 | 9.999 |
| 100 | 0.0000726 | 9.999 |
| Perp. | | 10.000 |

Therefore, if PV tends to 0 at perpetuity and if $n$ is infinite then given that:

$$PV\, £\, 1pa = \frac{1-PV}{i} \text{ then PV £ 1 pa in perpetuity will tend to } \frac{1-0}{i} = \frac{1}{i}$$

At a rate of 10% the PV £1pa in perp.$= \dfrac{1}{0.10} = 10$

---

**Example 2.9**

A freehold property produces a net income (annual net rent) of £15,000 a year. If an investor requires a return of 8%, what price should be paid? The income is perpetual so a PV £1 pa in perpetuity should be used.

| Income | £15,000 | |
|---|---|---|
| PV £1 pa in perp. at 8% | 12.5 | $(1/0.08 = 12.5)$ |
| Estimated price | £187,500 | |

It should be noted that £15,000/0.08 = £187,500 which is a commonly used way to set out an income in perpetuity PV calculation.

## Present Value of £1 pa in perpetuity deferred n years (PV £1 pa in perp defd or YP of a reversion to perpetuity)

A further common valuation application of the PV of £1 pa is known as the PV of £1 pa in perpetuity deferred ('years' purchase of a reversion to perpetuity'). It shows the PV of the right to receive a perpetual income starting at a future date and is found by multiplying PV £1 pa in perpetuity by the PV of £1 at the same rate of interest for the period of time that has to pass before the perpetual income is due to commence. It is useful to property valuers, as often property will be assumed to revert to a perpetual higher income after an initial period of time, such as following a rent review or renewal of a lease after a period of under renting.

$$PV\, £\, 1 \text{ pa perp deferred} = PV\, £\, 1 \text{ pa in perp} \times PV\, £\, 1 = \frac{1}{i} \times \frac{1}{(1+i)^n}$$

---

**Example 2.10**

Calculate in two ways the PV of a perpetual income of £600 pa beginning in seven years' time using a discount rate of 12%:

| Income | £600 |
|---|---|
| PV £1 pa in perp at 12% | 8.33 |
| Capital value (CV) | £5,000 |
| PV £1 in 7 years at 12% | 0.45235 |
| PV | £2,262 |

or

| Income | £600 |
|---|---|
| PV £1 pa perp def'd 7 years at 12% | 3.7695* |
| PV | £2,262 |

*Note that 8.33 × 0.45235 = 3.7695

The answer can also be found by deducting the PV £1 pa for seven years at 12% from the PV £1 pa in perpetuity at 12% 8.3334 – 4.5638 =3.7695.

The use of this PV or discounting technique to assess the price to be paid for an investment or the value of an income-producing property ensures the correct relationship between future benefits and present worth; namely that the investor will obtain both a return on capital and a return of capital at the market-derived rate of interest used (yield or discount rate) in the calculation.

This last point is important and can be missed when PVs are added together to produce the PV of £1 pa and labelled 'Years' Purchase' by the UK property valuation professional; the emphasis is on the UK, as few valuers in other countries use this old term which invariably has to be translated for non-UK clients into PV of £1 pa.

### Example 2.11

An investor is offered five separate investment opportunities on five separate occasions; each will produce a certain cash benefit of £10,000, the first in exactly one year's time and the other four at subsequent yearly intervals. The investor is seeking a 10% return from his money. What price should be paid for each investment?

| | Year 1 | Year 2 | Year 3 | Year 4 | Year 5 |
|---|---|---|---|---|---|
| Benefit | £10,000 | £10,000 | £10,000 | £10,000 | £10,000 |
| PV £1 at 10% | 0.9090 | 0.8264 | 0.7513 | 0.6830 | 0.6209 |
| PV (investment price today) | £9,090 | £8,264 | £7,513 | £6,830 | £6,209 |
| Amount of £1 at 10% | 1.1000 | 1.2100 | 1.3310 | 1.4641 | 1.6105 |
| | £10,000 | £10,000 | £10,000 | £10,000 | £10,000 |

The figures of PV – £9,090, £8,264, £7,513, £6,830, £6,209 – show the individual prices to be paid today for each investment in order that the investor can achieve a 10% return on capital invested and obtain the return of the capital invested. For example, £7,513 is paid for a future benefit of £10,000 in three years time, in three years time the receipt of £10,000 repays the investment of £7,513; the difference represents the equivalent compound interest at 10% on the £7,513. This is what is meant by a 10% return.

In other words, the investor is exchanging a sum of money today for a known future sum which will be equal to the sum of money today, plus the interest forgone if the capital had been invested elsewhere at the same rate of interest. In each case the receipt of £10,000 at the due date returns the respective capital sum or price paid and the difference between the price today and the £10,000 is equivalent to the 10% annual compound interest that would have been earned on the purchase sums if they had been saved at the investor's 10% required rate of return.

**Example 2.12**

If an investor is offered a single investment which generates a 'certain' cash benefit of £10,000 at the end of each year for the next five years and wishes to achieve a 10% return, what price should they pay today?

The answer can be found by adding the five sums together, e.g. £9,090 + £8,264 + £7,513 + £6,830 + £6,209 which equals a sum of £37,908. However, it is quicker to multiply the £10,000 a year by the PV £1 pa for five years at 10%.

£10,000 × 3.7908 = £37,908

Again, the investor must achieve a return of capital, (the initial £37,908) and a return of 10% on the capital (that is the interest forgone). The proof is shown in the table below:

| Year | Capital Outstanding | Return at 10% | Income | Return of Capital (Balance) |
|------|--------------------|--------------|--------|-----------------------------|
| 1 | 37,908.00 | 3,790.80 | 10,000 | 6,209.20 |
| 2 | 31,698.80 | 3,169.88 | 10,000 | 6,830.12 |
| 3 | 24,868.68 | 2,486.87 | 10,000 | 7,513.13 |
| 4 | 17,355.55 | 1,735.55 | 10,000 | 8,264.45 |
| 5 | 9,091.10 | 909.11 | 10,000 | 9,090.89 |

Note: rounding errors are due to calculation to two decimal places only.

## PV of £1 pa dual rate (YP dual rate)

The authors believe that the dual rate concept, which uses two separate rates, one being a speculative or risk rate to represent the market return on capital (the remunerative rate) and the other being a safe rate or sinking fund rate to provide for a return of capital through accumulations in a sinking fund, is no longer accepted by investors and so should not be used by valuers. (See Mackmin (2007) pp 80–95.) Its use is believed to be in decline. Appendix A contains a fuller description of the concept; only brief references are made to it in this edition.

It represents the PV of one pound per annum receivable at the end of each year after allowing for a sinking fund at a given rate of interest to replace the invested capital and for a return on capital at a market risk rate. It was a popular tool for assessing the PV of lease-hold interests in property up to the end of the twentieth century when its conceptual weaknesses were exposed by various research papers and journal papers presented by academics and practitioners. The formula is:

$$\frac{1}{i + ASF}$$

Here, $i$ is the remunerative rate or risk rate and ASF is the annual sinking fund which provides for replacement of capital at a low safe accumulative rate. When used by property valuers, $i$ was traditionally assumed to be 1% to 2% above the risk rate derived from sales of freehold interests in comparable properties. The sinking fund was at a low savings rate which had to be guaranteed and so the non profits rate within assurance fund investments was adopted of 2% to 4%. Currently, this might be as low as 0.5% to 1.0%.

## Dual rate adjusted for tax

Concern over the affect of tax on the sinking fund accumulations and on that part of the income used for the annual sinking fund payments lead to the use of tax-adjusted dual rates for which the formula is:

$$\frac{1}{i+s.T_G} \quad where \ T_G = \frac{100}{100 - rate \ of \ tax}$$

### Example 2.13

Value a terminating income of £1,000 for 10 years on a dual rate basis a) on a gross basis and b) on a tax adjusted basis, demonstrate that each achieves the objectives of the calculation. Assume a risk rate of 10% and a safe rate of 2.5% net of tax, tax is to be taken at 50%.

| a) Income | £1,000 |
|---|---|
| PV £1 pa for 10 years at 10% and 2.5% | 5.2838 |
| PV dual rate | £5,283.80 |

Proof
A 10% return on £5,283.80 is £528.38 which leaves an annual sum for reinvesting in a sinking fund of £471.62 and so:

| ASF | £471.62 |
|---|---|
| Amount of £1 pa for 10 years at 2.5% | 11.2034 |
| ASF in 10 years | £5,283.75 |

| b) Income | £1,000 |
|---|---|
| PV £1 pa for 10 years at 10% and 2.5% adj tax at 50% | 3.5904 |
| Present Value dual rate adj tax | £3,590.40 |

Proof:
A 10% return on £3,590.40 is £359.04 which leaves £640.96 available for tax at 50% and for reinvestment of the balance in the ASF and so:

| Amount for ASF | £640.96 |
|---|---|
| Less tax at 50% | £320.48 |
| Net amount for ASF | £320.48 |
| Amount of £1 pa for 10 years at 2.5% | 11.2034 |
| ASF in 10 years | £3,590.47 |

Small variations are due to working to four decimal places.

## Annuity £1 will purchase (A £1 wp)

This function shows the amount that will be paid back at the end of each year for $n$ years at $i\%$, in return for £1 invested. It calculates what is known as the annuity that £1 will purchase. It can be used to calculate the annual equivalent of a capital sum, although valuers tend to prefer to divide a capital sum by the PV of £1 pa to calculate an annual equivalent.

An annuity entitles the investor to a series of equal annual sums. These sums may be perpetual or for a limited number of years. When money is invested in a savings account, interest accumulates on the principal which remains in the account. When an annuity is purchased, however, the initial purchase price is lost forever to the buyer. The return from the savings account is all annual interest on capital. The return from an annuity represents part interest on capital and part the return of the purchase price.

These constituent parts will be referred to as return on capital (interest) and return of capital.

In an annuity, the original capital outlay must be returned by the end of the investment. If not, how could the rate of interest earned by an annuity investment be compared with the rate of interest earned in a building society account?

The amount of an annuity will depend upon three factors: the purchase price, $i$ and $n$.

### Example 2.14

What annuity will £50 purchase for five years if a 10% yield is required by the purchaser?

| | |
|---|---|
| Purchase sum | £50 |
| Annuity £1 wp for 5 years at 10% | 0.2638 |
| Annuity | £13.19 |

The £13.19 has two constituent parts. The return on capital is 10% of the outlay, i.e. £5 pa. This leaves £8.19 extra. This is the return of capital. However, 5 × £8.19 does not return £50 because each payment is in the nature of a sinking fund instalment and assumed to be earning interest, in this case at 10%.
Proof:

| | |
|---|---|
| ASF | £8.19 |
| A £1 pa for 5 years at 10% | 6.1051 |
| CV | £50 |

Example 2.14 shows that the return *on* capital and return *of* capital are both achieved, as a sinking fund is inherent in the annuity.

It can be noted that if the return on capital is $i$ and the return of capital is SF, then the formula for the annuity £1 wp must be $i + SF$, where $i$ is the rate of interest required and SF is the ASF required to replace the capital outlay in $n$ years at $i\%$.

## Example 2.15

Calculate the annuity £1 will purchase for 10 years at 10%, if £6.1446 will purchase £1 at the end of each year for 10 years.

£1 will purchase $\dfrac{1}{6.1466} = £0.1627$

Therefore, the annuity £1 will purchase for 10 years at 10% is £0.1627.

The annuity factor can also be calculated from the formula $i + SF$. As the sinking fund factor is the annual factor of £1 needed to build up the return of capital, this is known technically as to amortise £1, so the total partial payment required for recovery of capital and for interest on capital must be the amortisation factor plus the interest rate.

Therefore, the annuity £1 will purchase for 10 years at 10% is:

|  |  |
|---|---|
|  | $i = 0.1000$ |
| Plus the ASF to replace £1 in 10 years at 10% + ASF | 0.0627 |
|  | £0.1627 |

An annuity generally means a life annuity, a policy issued by a life assurance company where the investor will receive, for the rest of his or her life, a given annual income in exchange for a given capital sum. The calculations undertaken by the life office's actuaries have to take into account many factors, including life expectancy, which is determined by lifestyle. This is beyond the scope of this book, which is concerned with annuities related to property where the period of time over which the annuity will be paid is certain. Before proceeding with further consideration and analysis of annuities, there are a number of common terms which need to be explained.

### Annuity terminology

*In arrears:* This means that the first annual sum will be paid (received) 12 months after the purchase or taking out of the policy. The payments could be weekly, monthly, quarterly or for any period provided they are in arrears.

*In advance:* This means that the payments are made at the beginning of the week, month, year, etc.

*An immediate annuity:* The word 'immediate' is used to distinguish a normal annuity from a *deferred* annuity. An immediate annuity is one where the income commences immediately either 'in advance' or 'in arrears', whereas in the case of a deferred annuity capital is exchanged today for an annuity 'in advance' or 'in arrears', the first such payment being deferred for a given period of time longer than a year. In assurance terms one might purchase, at the age of 50, a life annuity to begin at the age of 65. This would be a deferred annuity.

The majority of published annuity tables are based on an 'in arrears' assumption.

The distinction between '*return on*' and '*return of*' capital is important in the case of life annuities because the capital element is held by HM Revenue & Customs (HMRC) to be the return of the annuitant's capital and therefore is exempt from income tax. In practice, HMRC has had to indicate how the distinction is to be made. Quite clearly, a precise distinction is not otherwise possible because of the uncertain nature of the annuitant's life.

The distinction in the case of certain property valuations undertaken on an annuity basis is important, because HMRC is not allowed to distinguish between 'return on' and 'return of' capital other than for life annuities. Thus, tax may be payable on the whole income from the property and the desired return may not be achieved unless this factor is accounted for in the valuation.

---

**Example 2.16**

What annuity will £4,918 purchase over six years at 6%?

|  |  |
|---|---|
|  | £4,918 |
| Annuity £1 wp for 6 years at 6% | 0.2033 |
|  | £1,000 pa |

---

**Example 2.17**

What annuity will £1,000 purchase over three years at 10%?

|  |  |
|---|---|
|  | £1,000 |
| Annuity £1 wp for 3 years at 10% | 0.4020 |
|  | £402 pa |

Prove that the investor recovers his or her capital in full and earns interest from year to year at 10%.

| Year | Capital outstanding | Interest at 10% | Income (annuity) | Return of Capital |
|---|---|---|---|---|
| 1 | £1,000 | £100 | £402 | £302 (£402 – £100) |
| 2 | £698 (1,000 – 302) | £69.80 (10% on £698) | £402 | £332.20 (£402 – £69.80) |
| 3 | £365.80 (698 – 32.20) | £36.58 (10% on £365.80) | £402 | £365.42 (£402 – £36.58) + £999.42 (error due to rounding) |

---

**Example 2.18**

Calculate the capital cost of an annuity of £500 for six years due in advance at 10%.

|  |  |  |
|---|---|---|
| Annuity income |  | £500 |
| PV £1 pa for 5 years ($n - 1$) at 10% | 3.79 |  |
| plus | 1.0 | 4.79 |
|  |  | £2,395 |
| or |  | £500 |
| PV £1 pa for 6 years at 10% | 4.3553 |  |
|  | X 1.1 | 4.79 |
|  |  | £2,395 |

or £500 in advance for six years is £500 in arrears for five years plus an immediate £500:

|  | £500 |
| PV £1 pa for 5 years at 10% | 3.79 |
|  | £1,895 |
|  | +£500 |
|  | £2,395 |

Annuities may be variable. The present worth of variable income flows could be expressed as

$$V = \frac{1_i}{PV_1} + \frac{1_2}{PV_2} + \frac{1_3}{PV_3} \ldots .. \frac{1_n}{PV_n}$$

Whilst it is possible to have a variable annuity changing from year to year, it is more common to find the annuity changing at fixed intervals of time.

---

**Example 2.19**

How much would it cost today to purchase an annuity of £1,000 for five years followed by an annuity of £1,200 for five years on a 10% basis?

| Immediate annuity |  | £1,000 |  |
| PV £1 pa for 5 years at 10% |  | 3.79 | £3,790 |
| Plus deferred annuity |  | £1,200 |  |
| PV £1 pa for 5 years at 10% | 3.79 |  |  |
| PV £1 for 5 years at 10% | 0.62 | 2.3498 | £2,820 |
|  |  |  | £6,610 |

It will be shown later that this approach is the same as that used by valuers when valuing a variable income flow from a property investment.

In the formula $i$ + SF, $i$ will remain the same whatever the length of the annuity. However, the value of each year of the annuity will reduce as time increases due to the effect of SF. For a perpetual annuity, SF will be infinitely small, and tends towards 0. A perpetual annuity therefore = $i$.

---

**Example 2.20**

What perpetual annuity can be bought for £1,500 if a 12% return is required by the investor?

| Invested sum | £1,500 |
| ($i = 0.12$) |  |
| Annuity £1 wp at 12% in perp. | 0.12 |
|  | £180 pa. |

The formula for a perpetual annuity is $i$; the formula for a YP in perpetuity is $1/i$ and is therefore the reciprocal. The annuity £1 will purchase is the reciprocal of the present value of £1 pa and it follows that there must be a second formula for PV £1 pa namely:

$\dfrac{1}{i+SF}$ as the limited term annuity formula is $i + SF$.

This is easily proved.

Annuity £1 wp $= i + SF$:

Hence capital x $(i + SF)$ = income and income x PV £1 pa must = Capital.

Therefore PV £ 1pa $= \dfrac{1}{i+SF}$

## Example 2.21

How much should A pay for the right to receive £1 pa for 25 years at 10%?

$$PV£\,1pa = \frac{1}{i+SF} = \frac{1}{i + \dfrac{i}{(1+i)^n - 1}} = \frac{1}{0.10 + \dfrac{0.10}{(1+0.10)^{25} - 1}}$$

$$= \frac{1}{0.10 + 0.010158} = 9.077$$

If the PV of £1 pa is 9.077, then A should pay £9.077.

Check

$$PV£\,1pa = \frac{1-PV}{i} = \frac{1 - 0.092296}{0.10} = 9.077$$

The function of SF in this PV £1 pa formula is exactly the same as that of PV in the original – that is, to reduce the value of the PV £1 pa figure as $n$ decreases. The PV £1 pa figure must include a sinking fund element as the valuation of an investment involves equating the outlay with the income and the required yield so that both a return on capital and a return of capital are received.

## Example 2.22

Value an income of £804.21 pa receivable for three years at a required yield of 10%. Illustrate how a return on and a return of capital are achieved.

| | |
|---|---|
| Income | £804.21 |
| PV £1 pa for 3 years at 10% | 2.4869 |
| | £2,000 |

The return *on* and a return *of* capital can be demonstrated by constructing a table:

| Year | Capital outstanding | Interest on capital outstanding | Income | Return of Capital |
|---|---|---|---|---|
| 1 | £2,000 | £200 | £804.21 | £604.21 (£804.21 – £200) |
| 2 | £1,395.79 | £139.58 | £804.21 | £664.63 (£804.21 – £139.58) |
| 3 | £731.16 | £73.12 | £804.21 | £731.09 (£804.21 – £73.12 error due to rounding) |

A rate of return is received on outstanding capital (i.e. that capital which is at risk) only. The table assumes that some capital is returned at the end of every year so that the amount of capital outstanding is reduced year by year. Interest on capital therefore decreases and more of the fixed income is available to return the capital outstanding. The last column shows how the sinking fund accumulates. £604.21 is the first instalment: £664.63 represents the second instalment of £604.21 plus one year's interest at 10%. £731.09 represents the third and final sinking fund instalment of £604.21 plus one year's interest at 10% on £1,268.84 (£664.63 + £604.21).

The three return of capital figures summate to the original outlay of £2,000.

The above type of table is also used to show how a mortgage is repaid.

# Mortgages

When a property purchaser borrows money by way of a legal mortgage, it is usually agreed between the parties that the loan will be repaid in full by a given date in the future. Like anyone else lending money, the building society or bank, known as the mortgagee, will require the capital sum to be repaid and will require interest on any outstanding amounts of the loan until such time as the loan and all interest are recovered. This is comparable to the purchase of an annuity certain. Indeed, conceptually, from the point of view of the mortgagee, it is the purchase of an annuity. It follows that the mortgagee will require a return on capital and the return of capital.

The repayment mortgage allows the mortgagor (borrower) to pay back a regular sum each year, often by equal monthly instalments. This sum is made up of interest on the capital owed and the partial return of capital. In the early years of the mortgage most of the payment represents interest on the loan outstanding and only a small amount of capital is repaid. However, over the period of the loan, as more and more capital is repaid, the interest element becomes smaller and the capital repaid larger.

For the purpose of explanation, interest is assumed fixed throughout the term of the loan. In practice, lenders offer a wide range of mortgage deals but the underlying concept is the same – a requirement that capital borrowed will be repaid and that interest will be paid on money owed. The rates of interest and pattern of payment can be infinitely variable and often involve lower interest payments during the early years.

The valuer who can understand the concept of the normal repayment mortgage and can solve standard problems that face mortgagors and mortgagees should readily understand most investment valuation problems. The following examples are indicative of such mortgage problems.

**Example 2.23**

The sum of £10,000 has been borrowed on a repayment mortgage at 6% for 25 years.

(i)   Calculate the annual repayment of interest and capital.

(ii)  Calculate the amount of interest due in the first and tenth years.

(iii) Calculate the amount of capital repaid in the first and tenth years.

The formula for the annuity £1 will purchase is $i + SF$.

Its reciprocal, the PV £1 pa is $\dfrac{1}{i + SF}$ or $\dfrac{1 - PV}{i}$

The mortgagee is effectively buying an annuity. There is in effect an exchange of £10,000 for an annual sum (annuity) over 25 years at 6%. What is the annual sum?

| | |
|---|---|
| Mortgage sum | £10,000 |
| Annuity £1 wp for 25 years at 6% | 0.07823 |
| Annual repayment | £782.30 |

Or £10,000 divided by PV £1 pa for 25 years at 6% which is £10,000/12.7834 = £782.30.

As interest in the first year is added immediately to capital borrowed, then:

£10,000 × 0.06 = £600 interest due in first year.

As the total to be paid is £782.30 so the amount of capital repaid will be £782.30 − £600 which is £182.30 in the first year.

The amount of interest in the tenth year will depend upon the amount of capital outstanding at the beginning of the tenth year, i.e. after the ninth annual payment.

Although a mortgage calculation is on a single rate basis, as the formula is $i + SF$, one can assume that £182.30 a year is a notional sinking fund accumulating over nine years at 6%.

| | |
|---|---|
| | £182.30 |
| Amount of £1 pa for 9 years at 6% | 11.4913 |
| Capital repaid is | £2,094.86 |

Capital outstanding is £10,000 less £2,094.86 which is £7,905.14; or calculate the value at 6% of the right to receive 16 more payments of £782.30.

| | |
|---|---|
| | £782.30 |
| PV £1 pa for 16 years at 6% | 10.1059 |
| | £7,905.85 |

Capital outstanding is £7,905.85.

Therefore, interest will be £7,905.85 × 0.06 = £474.35 and capital repaid in year 10 is £782.30 − £474.35 which is £307.95.

In passing it may be observed that the capital repaid in year 1 plus compound interest at 6% will amount to £208 after nine years:

| | |
|---|---|
| Year 1 capital | £182.30 |
| Amount of £1 for 9 years at 6% | 1.6895 |
| Capital repaid in tenth year | £308 (small error due to rounding). |

As interest on a mortgage is based on capital outstanding from year to year, the change in capital repaid from year to year must be at 6%.

|  |  |
|---|---|
|  | £182.30 |
| Amount of £1 for 24 years at 6% | 4.0489 |
| Capital repaid in last year | £738.11 |

As the annual payment is £782.30, so £782.30 – £738.11 represents the interest due in the last year of the mortgage, namely £44.19.

In many cases the rate of interest will change during the mortgage term. If it does, the mortgagor will usually have the choice of varying the term or repaying at an adjusted rate.

---

### Example 2.24

Given the figures in Example 2.23, advise the borrower on the alternatives available if the interest rate goes down to 5% at the beginning of year 10.
*Either:* (a) continue to pay £782.30 a year and repay the mortgage earlier or (b) reduce the annual payment.

(a)  *No change in annual payment*

Amount of capital outstanding at beginning of year 10 (i.e. the end of year 9) is (as before) £7,905.85.
This will be repaid at an annual rate of £782.30.
£7,905.85 divided by £782.30 = PV £1 pa for *n* years at 5% =10.11.
Substituting in the equation or interpolation in the PV £1 pa tables at 5% gives *n* as between 14 and 15 years, i.e. the loan is repaid before the due date.

(b)  *Change in annual payment*

|  |  |
|---|---|
| Capital outstanding is | £7,905.85 |
| Annuity £1 wp for 16 years at 5% | 0.09227 |
|  | £729.47 |

A change in interest at the beginning of year 10 means that, including the payment in year 10, there are 16 more payments due. The capital outstanding multiplied by the annuity £1 will purchase for 16 years, or divided by PV £1 pa for 16 years, at 5% (the new rate of interest) will give the new annual payment. This could be called the annual equivalent of £7,905.85 over 16 years at 5%.

# The interrelationship of the functions

A valuation student's knowledge of the interrelationship of the functions of £1 is sometimes tested by problems requiring the use of information provided by one particular table to calculate a related figure from another table. Knowledge of the formulae is essential. It can be seen that the amount of £1 $[(1 + i)^n$ or $(A)]$ is present in each formula:

$$1 \quad PV = \frac{1}{A}$$

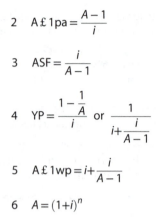

$$2 \quad A£1pa = \frac{A-1}{i}$$

$$3 \quad ASF = \frac{i}{A-1}$$

$$4 \quad YP = \frac{1-\frac{1}{A}}{i} \text{ or } \frac{1}{i+\frac{i}{A-1}}$$

$$5 \quad A£1wp = i+\frac{i}{A-1}$$

$$6 \quad A = (1+i)^n$$

It can also be seen from the above that the six functions fall into three sets of reciprocals: 1 and 6; 2 and 3; 4 and 5. The development of investment calculators, PCs and laptops means that any function can now be readily calculated without resource to transposition of formula, allowing students more opportunity to develop their appreciation of the use of the various financial functions. The following single example illustrates how knowledge of one function can be used to assess another function.

**Example 2.25**

Showing your workings, calculate to four decimal places:

(i)  PV £1 pa at 5% for 20 years given the PV £1 for 20 years at 5% is 0.37689.

$$PV = \frac{1}{A} \text{ and } PV£1pa = \frac{1-PV}{i}$$

PV £1 pa for 20 years at 5% $= \dfrac{1-0.37689}{0.05} = \dfrac{0.62311}{0.05} = 12.4622$

(ii)  ASF necessary to produce £1 after 15 years at 3% given that the Amount of £1 for 15 years at 3% is 1.558.

$$ASF = \frac{i}{A-1} = \frac{0.03}{1.558-1} = 0.0538$$

## Properties of exponents

In the compound interest formula $(1+i)^n$, $n$ is the exponent and the expression means that $(1+i)$ is multiplied by itself $n$ times. Exponents have certain properties and a knowledge of these can be useful when manipulating the various formulae.

1. Any number raised to the zero power equals 1.
2. A fractional exponent is the root of a number:

$$(1+i)^{1/2} = (1+i)^{0.5} = \sqrt[2]{(1+i)} \text{ similarly}$$

$$(1+i)^{1/5} = (1+i)^{0.2} = \sqrt[5]{(1+i)}$$

3. When a number being raised to a power is multiplied by itself, the exponents are added:

$$(1+i)^4 \times (1+i)^6 = (1+i)^{10}$$

and if being divided by itself they are subtracted

$$(1+i)^{10} \div (1+i)^5 = (1+i)^5$$

---

**Example 2.26**

Given that ASF 16 years at 4% is 0.04582 find the Annuity £1 will purchase for 14 years at 4%.

$$\frac{A\pounds 1 \ for \ 16 \ years \ at \ 4\%}{A\pounds 1 \ for \ 2 \ years \ at \ 4\%} = A\pounds 1 \ for \ 14 \ years \ at \ 4\%$$

$$ASF = \frac{i}{A-1} = 0.04582$$

$$A-1 = \frac{0.04}{0.04582} = 0.873$$

$$A = 0.0873 + 1 = 1.873$$

$$\frac{1.873}{(1+i)^2} = A\pounds 1 \ for \ 14 \ year \ at \ 4\%$$

$$\frac{1.873}{(1.04)^2} = \frac{1.873}{1.0816} = 1.7317$$

Thus, A £1 wp for 14 years at 4% $= i + \dfrac{i}{A-1}$ and now substituting from above the solution becomes:

$$= 0.04 + \frac{0.04}{1.7317-1} = 0.04 + \frac{0.04}{0.7317} = 0.04 + 0.05467 = 0.09467 \,.$$

Some valuation tutors still insist on testing this competence even further by using false figures in the set problems, thereby, in their view, fully testing their students' mathematical skills. The advent of financial calculators, Excel and valuation software packages, whilst not diminishing the need to understand the construction of the six financial functions of £1, have largely replaced the need for this level of appreciation of the interrelationships, with the opportunity to undertake more useful accurate application of the functions to situations where compounding or discounting is more properly required at intervals other than yearly. This step in the use of the six functions requires an awareness of the differences between nominal and effective rates of interest.

# Nominal and effective rates of interest

The rates of return provided by different forms of saving and from different forms of investment are usually compared by means of the annual rate of interest being paid or earned. For example, banks and building societies usually quote the annual rate of interest that will be earned in savings accounts, such as 10% per annum, representing the amount of interest that will be added to each £100 of savings left in the account for a full 12 months. Sometimes, however, it is necessary to check whether the rate of interest quoted is nominal or effective. The difference in annual interest can be significant depending on the frequency with which interest is added and at what rate.

For example, if a building society pays interest at 10% a year and it is added at the end of each complete year with the affect that £10 is added to the account for every £100 held for a year in that account, then the rate of 10% is both the nominal rate of interest and the effective rate of interest.Consider, on the other hand, building society B which pays 10% a year with interest paid half-yearly. 10% a year paid half-yearly at a rate of 5% every six months means a higher effective rate than the nominal 10% because the interest paid after six months will itself earn interest over the second half of the year. The total accumulation of £1 invested in an account for one year will therefore be £$(1.05)^2$ = £1.1025. Interest is 10.25 pence, accumulated on £1 invested. The effective rate of interest is therefore 10.25%.

However, if the 10% is quoted as the annual percentage rate (APR) or as an effective rate of interest but is actually credited every six months, then it is important to recognise that 5% or £5 for every £100 will not be added after the first six months. Instead, the savings provider will calculate the effective rate for every six months. So the amount earned every six months can be found by using the amount of £1 formula. The effective 10% becomes 4.88% payable every six months because:

$$(1+i)^n = (1+i)^2 = 1.10$$

The interest is to be paid twice a year, but after the interest is added, and after interest is added to the interest earned in the first six months, i.e. after compounding has occurred, the total per £100 is not to exceed 10% or 0.10 per £1. The problem is to calculate the six monthly rate of interest:

$$(1+i)^2 = 1.10$$
$$\therefore (1+i) = \sqrt[2]{1.10} = 1.0488$$
$$i = 1.0488 - 1 = 0.0488$$
$$0.0488 \times 100 = 4.88\%$$

If interest is to be paid quarterly, the rate of interest per quarter becomes 2.411%. This can be checked by substituting in the Amount of £1 formula:

$$(1+i)^4 = (1+0.02411)^4 = (1.02411)^4 = 1.10$$

Investments are therefore best compared by means of the annual effective rate of interest. Building societies, banks and all other finance companies are required to disclose this annual equivalent rate (AER) on savings and the APR on loans and mortgages. The APR may be based on sums which include arrangement costs.

This principle leads to the conclusion that given a specific effective rate of interest per year, the effective rate per period will get ever smaller as the interest earning period gets smaller. For many purposes a daily rate is used and interest is earned and credited on a daily basis. Similarly, certain loans and mortgages will have interest charged on the amount of the loan or debt outstanding from day to day or from month to month. This concept leads naturally to the possibility of continuous compounding.

## Continuous compounding

When interest is added more frequently than annually, the compound interest formula $(1 + i)^n$ is adjusted to:

$$\left(1 + \frac{i}{m}\right)^{mn}$$

Where $m$ is the number of times per year that interest is added, $i$ is the nominal rate of interest per year and $n$ is the number of years. Here, the more frequently interest is added, the greater the annual effective rate of interest will be.

If, for example, the nominal rate of interest is 12% per year and interest is added monthly then the effective rate is 12.68%:

$$\left(1 + \frac{0.12}{12}\right)^{12} = (1.01)^{12} = 1.1268$$

The greater the number of times in the year that interest is added, the greater the total sum at the end of the year will be, but there must be a limit, because whilst the number of periods becomes infinitely large, the rate of interest per period becomes infinitely small.

Potentially, $m$ might tend towards infinity. This would imply the immediate reinvestment of earned interest, or *continuous compounding*.

Mathematically, the maximum sum to which £1 could compound in one year at 100% pa is given by the following series:

$$1 + \frac{1}{1} + \frac{1}{1 \times 2} + \frac{1}{1 \times 2 \times 3} + \frac{1}{1 \times 2 \times 3 \times 4} \cdots \frac{1}{m}$$

This is a convergent series summating to six places, to 2.718282 or $e$.

The mathematical proof behind continuous compounding can be found elsewhere, for our purposes it can be accepted that the maximum sum to which £1 can accumulate over $n$ years with interest at $i$ and with continuous compounding becomes $e^{in}$.

---

**Example 2.27**

To what sum will £1 compound over two years at 10% per year nominal rate of interest assuming continuous compounding?

$$e^{in} = 2.71828^{(0.1)(2)}$$
$$e^{in} = 2.71828^{(0.2)}$$
$$e^{in} = 1.221$$

£1 accumulating at 10% per year with interest added annually would only compound to (1.10)2 or £1.21.

The PV of £1 $= \dfrac{1}{(1+i)^n} = (1+i)^{-n}$

It follows then that where interest is compounding continuously:
The PV of £1 $= e^{-in}$

The foregoing demonstrates that the rate of interest used in calculations of compounding and discounting must be the effective rate for the period.

## Incomes in advance and non-annual incomes

The issue here is how should a valuer deal with the valuation of incomes received in advance? The principle behind PV £1 pa and PV £1 is that the unit of income is received at the end of the year, but for most let properties the requirement is for rent to be paid quarterly in advance. In lower value areas and for residential property the expectation is that rent is paid monthly in advance. The Investment Property Forum has made strong pleas to the property market to adopt in advance practices, but whilst investment advice is frequently provided on the basis of the 'true equivalent yield' which reflects the precise timing of rental payments, most valuers and most valuations are undertaken on the basis of in arrears assumptions.

If one needs to move from in arrears to in advance, then common sense can be applied to derive the appropriate formula.

1. PV £1 normally assumes that the sum is to be received at the end of the year $n$. If the sum is received at the start of year $n$ instead this will coincide with the end of year $(n - 1)$.

   Thus, the PV of £1 receivable in advance:

   $$= \frac{1}{(1+i)^{n-1}}$$

2. Amount of £1 pa normally assumes that each £1 is invested at the end of each year. If this becomes the start of each year instead, an extra £1 will accumulate for $n$ years – but the £1 paid at the end of the $n$th year will now be paid at the start of year $n$. The series is now:

   $$(1+i)^n + (1+i)^{n-1} \ldots (1+i)^2 + (1+i)$$

   This can be summated to:

   $$\left[ \frac{(1+i)^{n+1} - 1}{i} \right] - 1$$

3. PV £1 pa usually assumes income to be received at the end of each year. However, if it comes in advance, the series will read:

   $$1 + \frac{1}{(1+i)} + \frac{1}{(1+i)^2} \ldots \frac{1}{(1+i)^{n-1}} = \frac{1 - \dfrac{1}{(1+i)^{n-1}}}{i} + 1$$

**Example 2.28**

Calculate PV £1 pa for six years at 10% in advance.

$$\frac{1 - \dfrac{1}{(1+0.10)^{6-1}}}{i} + 1 = \frac{1 - \dfrac{1}{1.6105}}{0.10} + 1 = 3.79 + 1 = 4.79$$

Alternatively, this could be given by PV £1 pa for five years at 10% + 1: (3.79 + 1) = 4.79, or by PV £1 pa for six years at 10% × (1 + i).

In this book, so far, $i$ has always represented an annual interest rate. However, the formulae can be used for alternative interest rate periods which are commonly used in property investment transactions. It must be ensured, however, that $i$ and $n$ relate to the same time period.

The time period can be anything; so for example if the interest rate $i$ is quarterly, the time period $n$ must represent quarterly periods, i.e. for one year $n$ would equal four.

**Example 2.29**

Calculate the amount of £1 in seven years, six months at an interest rate of 3% per half year.

In this example, the interest rate $i$ is per half year, so $n$ should represent periods of half a year.

$i = 0.03$          $n = 15$ periods

$A = (1 + i)^n$

$= (1 + 0.03)^{15}$

$= 1.557967$

The same logic may be applied to all six functions as specified at the beginning of this chapter.

# Incomes receivable quarterly or monthly in advance

It has been noted that almost all property is let on quarterly or monthly in advance terms and, further, that although the market maintains the tradition of valuing on an in arrears assumption, and of comparing market capitalisation rates on the same basis, it is a simple process, if mathematically more trying, to find the PV on a quarterly or monthly basis.

*Parry's Valuation Tables* contain quarterly in advance figures and most valuation software includes options to switch assumptions. However, to do so can be dangerous if the in arrears rates are simply substituted in the formula to assess in advance PVs.

The formula for the PV £1 pa to be applied to an income quarterly in advance, single rate, is:

$$PV\,£1\ pa = \frac{1 - \left[\dfrac{1}{(1+i)^{(n)}}\right]}{4\left[1 - \dfrac{1}{\sqrt[4]{(1+i)}}\right]}$$

where $i$ is the annual effective rate of interest.

For example, the single rate PV £1 pa quarterly in advance, at an annual effective rate of 10% for 20 years, is 9.0393, compared with the equivalent annual in arrears YP of 8.5136, reflecting the advantages of receiving income both earlier and more regularly.

For monthly in advance, the formula becomes:

$$PV£1\ pa\ mia = \frac{1 - \left[\dfrac{1}{(1+i)^n}\right]}{12\left[1 - \dfrac{1}{\sqrt[12]{(1+i)}}\right]}$$

This edition does not recommend the use of dual rate for leasehold valuations, however, out of interest, the formula for the YP to be applied to an income receivable quarterly in advance, dual rate with a tax adjustment is:

$$PV£1pa = \frac{1}{4\left[1 - \dfrac{1}{\sqrt[4]{(1+i)}}\right] + \dfrac{4\left[1 - \dfrac{1}{\sqrt[4]{(1+s)}}\right]}{[(1+s)^n - 1] \times [1-t]}}$$

where $i$ is the annual effective remunerative rate, $s$ is the annual effective accumulative rate, and $t$ is the tax rate.

For example, the dual rate PV £1 pa quarterly in advance, at an annual effective remunerative rate of 10%, an annual effective accumulative rate of 3%, adjusted for tax at 40% for 20 years, is 6.3555, compared with the annual in arrears equivalent figure of 6.1718.

This formula in itself is probably sufficient to deter anyone from using dual rate quarterly or monthly in advance.

## Health warning

Switching to quarterly or monthly without amending $i$% to reflect the change will produce different opinions of PV. This would be strictly true if the comparison was genuinely with an income in arrears. However, the rates used by valuers, whilst derived from sales on the assumption of annual in arrears and applied in the same way, are known to be the market yields on investments where the income is actually paid quarterly or monthly in advance. To find the PV of a sum or PV of a quarterly or monthly sum payable in advance, one needs $i$ to be derived from comparable quarterly or monthly sales analysis. *It is dangerous to simply switch* as it can imply a higher value where no higher value would be achieved in the market.

## Summary

This chapter has explored the six single rate investment functions which form the most common basis of property valuations by the income approach and has indicated how the functions are used to assess both returns on various types of investments and borrowing by way of mortgages. The six basic functions and their associated formulae are set out below, together with those for in advance functions.

The amount of £1 – used to calculate compound interest:

$$A = (1+i)^n$$

The amount of £1 per annum – used to calculate the future worth of regular periodic investment:

$$A £ 1pa = \frac{(1+i)^n - 1}{i}$$

ASF – used to calculate the sum that must be invested annually to cover a known expense in the future, at a known date:

$$ASF = \frac{i}{(1+i)^n - 1} = \frac{i}{A - 1}$$

Present value of £1 – used to calculate the PV of sums to be received in the future:

$$PV £ 1 = \frac{1}{A} = \frac{1}{(1+i)^n}$$

PV of £1 per annum – also known as the YP and used to calculate the PV of a series of payments:

$$\frac{1 - PV}{i} = \frac{1 - \dfrac{1}{(1+i)^n}}{i} \quad \text{OR} \quad \frac{1}{i + SF}$$

Annuity £1 will purchase – this is used to calculate the annual sum given in exchange for an initial amount of capital:

$$\text{Annuity } £1 \text{ wp} = i + SF$$

The PV of £1 receivable in advance – this is used to calculate PV where the sum is received at the start of the year:

$$= \frac{1}{(1+i)^{n-1}}$$

The amount of £1 per annum receivable in advance – this is used to calculate the future worth of regular periodic investments made at the start of each year:

$$= \left[ \frac{(1+i)^{n+1} - 1}{i} \right] - 1$$

PV of £1 pa or YP receivable in advance – this is used to calculate the present worth of a series of payments received in advance:

$$= \frac{1 - \dfrac{1}{(1+i)^{n-1}}}{i} + 1$$

Present value of £1 pa income receivable quarterly in advance:

$$= \frac{1 - \dfrac{1}{(1+i)^n}}{4\left[1 - \dfrac{1}{\sqrt[4]{(1+i)}}\right]}$$

# Spreadsheet user

Spreadsheets are particularly useful for exploring the financial functions and investment concepts outlined in this chapter. These pages are designed for current but introductory users of Microsoft Excel to help them use the spreadsheet as both a valuation tool and a means of exploring and understanding the concepts of valuation more fully.

## Project 1

Excel can be used very effectively to generate your own set of valuation tables, which you can personalise to your own requirements, layout and style.

The example below shows how to construct a table for the Amount of £1 function. Only a small section of the spreadsheet is shown in the example: you should create the table for values of $i$ between, say, 2% and 25% at 0.5% intervals and for 100 years. This project demonstrates the power of spreadsheets to repeat a calculation from a simple copy command. If you generate the whole table as indicated above the Amount of £1 calculation will be performed 4,700 times!

The following list gives advice on how to construct a spreadsheet:

- Plan the table carefully first, decide on its layout and presentation.
- Consider the logical sequence of the calculation and the parentheses required in the formula.
- You must fully understand the nature of relative and absolute cell references in order to understand how the single copy command generates the table by looking up the appropriate values of $i$ and $n$ in column A and row 2.
- The symbol ^ is used to indicate to the power of.
- The table will require substantial compression to print on a single page. This is simple in Excel by clicking on the option to fit a single page in the print dialogue box.
- In the example below, the value of $i$ is divided by 100 to express the value as a percentage, as required in all valuation formulae. You could alternatively format the cells in row 2 as a percentage and enter the values as 0.02, 0.025, etc. The cell will be displayed as 2%, 2.5%, etc. and you can remove the/100 from the formula.

The following table shows the start of the Amount of £1 table. The whole table can be generated by copying the formula in cell C2 as a block for the whole table starting at cell C2 and extending for your chosen range of values of *i* and *n*.

Note: You only have to enter the formula in cell C4 and copy it both across and down to generate your table! The formula you enter is shown in bold, while the formula contained in the other cells are shown only to illustrate how the formula changes when it is copied.

| | A | B | C | D | E |
|---|---|---|---|---|---|
| 1 | | Interest (%) | | | |
| 2 | Years | 2 | 2.5 | 3 | 3.5 |
| 3 | | | | | |
| 4 | 1 | (1+(C$2/100))^$A4 | (1+(D$2/100))^$A4 | (1+(E$2/100))^$A4 | (1+(F$2/100))^$A4 |
| 5 | 2 | (1+(C$2/100))^$A5 | (1+(D$2/100))^$A5 | (1+(E$2/100))^$A5 | (1+(F$2/100))^$A5 |
| 6 | 3 | (1+(C$2/100))^$A6 | (1+(D$2/100))^$A6 | (1+(E$2/100))^$A6 | (1+(F$2/100))^$A6 |
| 7 | 4 | (1+(C$2/100))^$A7 | (1+(D$2/100))^$A7 | (1+(E$2/100))^$A7 | (1+(F$2/100))^$A7 |
| 8 | 5 | (1+(C$2/100))^$A8 | (1+(D$2/100))^$A8 | (1+(E$2/100))^$A8 | (1+(F$2/100))^$A8 |

The start of your Amount of £1 table should look like the following table:

The start of your Amount of £1 table spreadsheet, with cell C4 = (1+(C$2/100))^$A4

| | A | B | C | D | E | F | G | H | I | J |
|---|---|---|---|---|---|---|---|---|---|---|
| 1 | | | Interest (%) | | | | | | | |
| 2 | Years | | 2 | 2.5 | 3 | 3.5 | 4 | 4.5 | 5 | 5.5 |
| 3 | | | | | | | | | | |
| 4 | 1 | | 1.02 | 1.025 | 1.03 | 1.035 | 1.04 | 1.045 | 1.05 | 1.055 |
| 5 | 2 | | 1.0404 | 1.050625 | 1.0609 | 1.071225 | 1.0816 | 1.092025 | 1.1025 | 1.113025 |
| 6 | 3 | | 1.061208 | 1.076891 | 1.092727 | 1.108718 | 1.124864 | 1.141166 | 1.157625 | 1.174241 |
| 7 | 4 | | 1.08243216 | 1.103813 | 1.125509 | 1.147523 | 1.169859 | 1.192519 | 1.215506 | 1.238825 |
| 8 | 5 | | 1.104080803 | 1.131408 | 1.159274 | 1.187686 | 1.216653 | 1.246182 | 1.276282 | 1.30696 |
| 9 | 6 | | 1.126162419 | 1.159693 | 1.194052 | 1.229255 | 1.265319 | 1.30226 | 1.340096 | 1.378843 |
| 10 | 7 | | 1.148685668 | 1.188686 | 1.229874 | 1.272279 | 1.315932 | 1.360862 | 1.4071 | 1.454679 |
| 11 | 8 | | 1.171659381 | 1.218403 | 1.26677 | 1.316809 | 1.368569 | 1.422101 | 1.477455 | 1.534687 |
| 12 | 9 | | 1.195092569 | 1.248863 | 1.304773 | 1.362897 | 1.423312 | 1.486095 | 1.551328 | 1.619094 |
| 13 | 10 | | 1.21899442 | 1.280085 | 1.343916 | 1.410599 | 1.480244 | 1.552969 | 1.626895 | 1.708144 |
| 14 | | | | | | | | | | |

## Project 2

Construct a *ready reckoner* which displays the various values of £1 using the basic six functions of £1 and those additional functions that are most frequently used.

Create the spreadsheet so that it displays the values of £1 for any actual sums that might be required and in such a way that by changing the % figure all the sums are instantly recalculated, as the following spreadsheet shows:

|   | A | B | C | D | E |
|---|---|---|---|---|---|
| 7 |  |  |  |  |  |
| 8 | **INVESTMENT FUNCTIONS** |  | **Factor per £** | **Sum B6 x C9 to C26** |  |
| 9 | Amount £ |  | 2.59374246 | £2,593.74 |  |
| 10 | Amount £ pa |  | 15.9374246 | £15,937.42 |  |
| 11 | Amount £ pa in advance |  | 17.53116706 | £17,531.17 |  |
| 12 | Annual Sinking Fuind |  | 0.062745395 | £62.75 |  |
| 13 | Present Value of £ |  | 0.385543289 | £385.54 |  |
| 14 | Present Value of £pa(YP) |  | 6.144567106 | £6,144.57 |  |
| 15 | Present Value of £ per quarter in advance (YP) |  | 6.524027975 | £6,524.03 |  |
| 16 | Annuity £ will purchase |  | 0.162745395 | £162.75 |  |
| 17 | Mortgage repayment pa |  | 0.162745395 | £162.75 |  |
| 18 |  |  |  |  |  |
| 19 | YP in perpetuity |  | 10 | £10,000.00 |  |
| 20 | YP in perpetuity in advance |  | 11 | £11,000.00 |  |
| 21 | YP quarterly in advance in perpetuity |  | 10.61755509 | £10,617.56 |  |
| 22 | YP perpetuity deferred |  | 3.855432894 | £3,855.43 |  |
| 23 | YP quarterly in advance in perp.deferred |  | 4.093527115 | £4,093.53 |  |
| 24 | YP monthly in advance |  | 6.472552122 | £6,472.55 |  |
| 25 | Mortgage repayment pa (monthly basis) |  | 0.154498563 | £154.50 |  |
| 26 | YP Dual rate at 4% unadj |  | 5.455806907 | £5,455.81 |  |
| 27 | YP Dual rate at 4% adj.tax at 40% |  | 4.187236169 | £4,187.24 |  |
| 28 |  |  |  |  |  |
| 29 | The amount will be the amount saved, borrowed, to be accumulated to in a sinking fund, the annual rent etc. |  |  |  |  |
| 30 | The years will in some instances be the period of deferment eg YP perp deferred. |  |  |  |  |
| 31 |  |  |  |  |  |

# Chapter 3

## Discounted Cash Flow

## Introduction

Discounted cash flow (DCF) is an aid to the valuation or analysis of any investment producing a cash flow. In its general form, it has two standard products – net present value (NPV) and internal rate of return (IRR).

## NPV

Future net benefits receivable from the investment are discounted at a given 'target rate'. The sum of the discounted benefits is found and the initial cost of the investment deducted from this sum, to leave what is termed the net present value (NPV) of the investment, which may be positive or negative. A positive NPV implies that a rate of return greater than the target rate is being yielded by the investment; a negative NPV implies that the yield is lower than the target rate. The target rate is the minimum rate the investor requires in order to make the investment worthwhile, taking into account the risk involved and all other relevant factors. It will be governed in particular by the investor's cost of capital. The target rate may be called the hurdle rate or discount rate.

Investments may require the initial outlay of a large capital sum and investors will often borrow such sums of money to complete their purchases. The interest to be paid on that loan will be the investor's cost of capital at that time. It is clear that the rate of return from an investment where the initial capital has been borrowed should be at least equal to the cost of capital or a loss will result.

An alternative way of looking at this is that investors will always have alternative opportunities for the investment of their capital. Money may be lent quite easily to earn a rate of interest based on the cost of capital. The return from any investment should therefore compare favourably with the opportunity cost of the funds employed and this will usually be related to the cost of capital.

For these reasons, the target rate should compare well with the cost of capital. From this basis, a positive or negative NPV will be the result of the analysis and upon this result an investment decision can be made.

## Example 3.1

Find the NPV of the following project using a target rate of 13%.

**Outlay £10,000**

|  | Income | PV £1 at 13% | Discounted sum £ |
|---|---|---|---|
| Returns in year 1 | 5,000 | 0.8849 | 4,425 |
| Returns in year 2 | 4,000 | 0.7831 | 3,133 |
| Returns in year 3 | 6,000 | 0.6930 | 4,158 |
|  |  |  | £11,716 |
|  |  | Less outlay | £10,000 |
|  |  | NPV | £1,716 |

The investment yields a return of 13% and, in addition, a positive NPV of £1,716. In the absence of other choices, this investment may be accepted because the return exceeds the target rate. However, it is more usual for the investment decision to be one which requires a choice to be made between two or more investments.

## Example 3.2

An investor has £1,400 to invest and has a choice between investment A and B. The following returns are anticipated:

| Income flow | A | B |
|---|---|---|
|  | £ | £ |
| Year 1 | 600 | 200 |
| 2 | 400 | 400 |
| 3 | 200 | 400 |
| 4 | 400 | 600 |
| 5 | 400 | 600 |

The investor's target rate for both investments is 10%, the investments are mutually exclusive (only one of the two can be undertaken); which investment should be chosen?

**Income flow A**

|  |  | Income | PV £1 at 10% | Discounted sum |
|---|---|---|---|---|
| Year | 1 | 600 | 0.9091 | 545 |
|  | 2 | 400 | 0.8264 | 330 |
|  | 3 | 200 | 0.7513 | 150 |
|  | 4 | 400 | 0.6830 | 273 |
|  | 5 | 400 | 0.6209 | 248 |
|  |  |  |  | £1,546 |
|  |  |  | Less outlay | £1,400 |
|  |  |  | NPV | £146 |

**Income flow B**

|  |  | Income | PV £1 at 10% | Discounted sum |
|---|---|---|---|---|
| Year | 1 | 200 | 0.9091 | 182 |
|  | 2 | 400 | 0.8264 | 330 |
|  | 3 | 400 | 0.7513 | 300 |
|  | 4 | 600 | 0.6830 | 409 |
|  | 5 | 600 | 0.6209 | 373 |
|  |  |  |  | £1,594 |
|  |  |  | *Less outlay* | £1,400 |
|  |  |  | NPV | £194 |

From this information, investment B should be chosen. Each investment gives a return of 10% plus a positive NPV; B produces the greater NPV of £194 compared to A at £146.

If either investment had been considered to be subject to more risk, this factor should have been reflected in the choice of target rate.

In Example 3.2, both investments involved the same amount of capital or initial outlay. However, this will not always be the case, and, when outlays on mutually exclusive investments differ, the investment decision will not be so simple. An NPV of £200 from an investment costing £250 is considerably more attractive than a similar NPV produced from a £25,000 outlay.

However, how can this be reflected in an analysis? A possible approach is to express the NPV as a percentage of the outlay.

**Example 3.3**

Which of these mutually exclusive investments should be undertaken when the investor's target rate is 10%?

|  | Investment A<br>Outlay £5,000 | Investment B<br>Outlay £7,000 |
|---|---|---|
| *Income flow* |  |  |
| Year 1 | £3,000 | £4,000 |
| Year 2 | £2,000 | £3,000 |
| Year 3 | £1,500 | £2,000 |

**Income flow A**

|  |  | £ | PV £1 at 10% | Discounted sum |
|---|---|---|---|---|
| Year | 1 | 3,000 | 0.9091 | 2,727 |
|  | 2 | 2,000 | 0.8264 | 1,653 |
|  | 3 | 1,500 | 0.7513 | 1,127 |
|  |  |  |  | £5,507 |
|  |  |  | Less outlay | £5,000 |
|  |  |  | NPV | £507 |

**Income flow B**

| Year | | £ | PV £1 at 10% | Discounted sum |
|---|---|---|---|---|
| Year | 1 | 4,000 | 0.9091 | 3,636 |
| | 2 | 3,000 | 0.8264 | 2,479 |
| | 3 | 2,000 | 0.7513 | 1,503 |
| | | | | £7,618 |
| | | | Less outlay | £7,000 |
| | | | NPV | £618 |

At first sight, B might appear to be more profitable, but if the NPV is expressed as a percentage of outlay the picture changes.

$$A = \frac{£\,507}{£\,5,000} \times 100 = 10.14\,\%$$

$$B = \frac{£\,618}{£\,7,000} \times 100 = 8.33\,\%$$

On this basis A, and not B, should be chosen.

The NPV method is a satisfactory aid in the great majority of investment problems but suffers from one particular disadvantage. The return provided by an investment is expressed in two parts, a rate of return and a cash sum in addition, which represents an extra return. These two parts are expressed in different units which may make certain investments difficult to compare.

This problem is not present in the following method of expressing the results of a DCF analysis.

# IRR

The IRR is the discount rate which equates the discounted flow of future benefits with the initial outlay. It produces an NPV of 0 and may be found by the use of various trial discount rates.

**Example 3.4**

Find the IRR of the following investment.

| Outlay | **£6,000:** | Returns | Year 1 | £1,024 |
|---|---|---|---|---|
| | | | Year 2 | £4,000 |
| | | | Year 3 | £3,000 |

**Trying 10%:**

| Year | | £ | PV £1 at 10% | Discounted sum |
|---|---|---|---|---|
| Year | 1 | 1,024 | 0.9091 | 931 |
| | 2 | 4,000 | 0.8264 | 3,306 |
| | 3 | 3,000 | 0.7513 | 2,253 |
| | | | | £6,490 |
| | | | Less outlay | £6,000 |
| | | | NPV | £490 |

At a trial rate of 10%, a positive NPV results. £490 is too high – an NPV of 0 is required. The trial rate must be too low as the future receipts need to be discounted to a greater extent.

**Trying 16%:**

| | | £ | PV £1 at 16% | Discounted sum |
|------|---|-------|--------------|----------------|
| Year | 1 | 1,024 | 0.8621 | 931 |
| | 2 | 4,000 | 0.7432 | 2,972 |
| | 3 | 3,000 | 0.6407 | 1,923 |
| | | | | £5,778 |
| | | | Less outlay | £6,000 |
| | | | NPV | - £222 |

This time the receipts have been discounted too much. A negative NPV is the result, so the trial rate is too high. The IRR must be between 10% and 16%.

**Trying 14%:**

| | | £ | PV £1 at 14% | Discounted sum |
|------|---|-------|--------------|----------------|
| Year | 1 | 1,024 | 0.8772 | 899 |
| | 2 | 4,000 | 0.7695 | 3,076 |
| | 3 | 3,000 | 0.6750 | 2,025 |
| | | | | £6,000 |
| | | | Less outlay | £6,000 |
| | | | NPV | £0,000 |

As the NPV is £0 in Example 3.4, the IRR must be 14%.

Calculation of the IRR by the use of trial rates will be difficult when the IRR does not happen to coincide with a round figure, as in the following illustration.

**Outlay £4,925**
**Trying 11%:**

| | | £ | PV £1 at 11% | Discounted sum |
|------|---|-------|--------------|----------------|
| Year | 1 | 2,000 | 0.9009 | 1,802 |
| | 2 | 2,000 | 0.8116 | 1,623 |
| | 3 | 2,000 | 0.7312 | 1,462 |
| | | | | £4,887 |
| | | | Less outlay | £4,925 |
| | | | NPV | -£38 |

The trial rate is too high:

**Trying 10%:**

| | | £ | PV £1 at 10% | Discounted sum |
|------|---|-------|--------------|----------------|
| Year | 1 | 2,000 | 0.9091 | 1,818 |
| | 2 | 2,000 | 0.8264 | 1,652 |
| | 3 | 2,000 | 0.7513 | 1,503 |
| | | | | £4,973 |
| | | | Less outlay | £4,925 |
| | | | NPV | £48 |

The IRR is therefore between 10% and 11%.

The continued use of trial rates to make a more accurate estimation of the IRR will therefore be impracticable. The analysis has shown that the IRR lies between 10% and 11% but the accuracy of such an analysis is limited.

Linear interpolation can be carried out by the use of the following formula, where the lower trial rate = *ltr* and the higher trial rate = *htr*. The formula simply assumes a straight-line relationship between the trial rates and the resulting NPVs:

$$IRR = ltr + \frac{NPVltr}{NPVltr + NPVhtr} \times (htr - ltr)$$

In this case the lower trial rate is 10%; the higher trial rate is 11%; the NPV at the lower trial rate is £48 and the NPV at the higher trial rate is –£38.

The difference in trial rates is 1%.

$$IRR = 10\% + \left(\frac{48}{48+38}\right) \times (11\% - 10\%)$$
$$= 10 + \left(\frac{48}{36} \times 1\right)$$
$$= 10 + 0.558$$
$$= 10.558\%$$

Linear interpolation will provide a satisfactory answer in the majority of cases. However, it must be borne in mind that the mathematical relationship here is geometric not arithmetic; it is not a straight line. Because of this, linear interpolation is inaccurate to a certain extent, and the result should not be expressed to too many decimal places.

An IRR of 10.56% will be sufficiently reliable and precise for most purposes. Such calculations, however, are more readily undertaken using an investment calculator or 'Goal Seek' on Microsoft Excel which provides a full solution, namely 10.560884981651%; the following screenshot shows goal seek to the nearest whole percent:

| | A | B | C | D | E | F |
|---|---|---|---|---|---|---|
| 1 | | | | | | |
| 2 | | | Target Rate | 11% | | |
| 3 | | | Outlay | £4,925 | | |
| 4 | | Period | Cash Flow | PV £1 at D2 | Discounted sum | |
| 5 | | 1 | £2,000 | 0.90447901 | £1,808.96 | |
| 6 | | 2 | £2,000 | 0.81808228 | £1,636.16 | |
| 7 | | 3 | £2,000 | 0.73993825 | £1,479.88 | |
| 8 | | Present Value | | | £4,925.00 | |
| 9 | | Less Outlay | | | £4,925 | |
| 10 | | NPV | | | -£0.00 | |
| 11 | | | | | | |

A normal DCF with a target rate is set up for the problem on Excel. On Microsoft Vista the procedure is as follows:

1. Click data tab on tool bar.
2. Click 'What-If Analysis'.
3. Click 'Goal Seek' in drop down menu.
4. Set NPV cell, in this case, E10, to the value of £0.
5. Insert target rate, here cell D3, in 'By changing cell'.
6. Click OK.
7. Move cursor to the target rate cell and the full IRR is displayed in the equation bar; alternatively, cell D2 can now be expanded to the required number of decimal places.

## Comparative use of NPV and IRR

When analysing a range of investments, NPV and IRR results may be compared with one another. In some cases conflicting results may arise and additional analysis may be needed to determine the best choice of investment.

**Example 3.5**

On the basis of NPV and IRR methods rank the following investments, using a rate of 10% in each case.

| Investment A Outlay £5,000 | | Investment B Outlay £10,000 | |
|---|---|---|---|
| Income: | £ | Income: | £ |
| Year 1 | 600 | Year 1 | 3,342 |
| Year 2 | 2,000 | Year 2 | 3,342 |
| Year 3 | 4,000 | Year 3 | 3,342 |
| Year 4 | 585 | Year 4 | 3,342 |

(a) NPV analysis using a trial rate 10%

**Investment A:**

| Year | £ | PV £1 at 10% | Discounted sum |
|---|---|---|---|
| 1 | 600 | 0.9091 | 545 |
| 2 | 2,000 | 0.8264 | 1,653 |
| 3 | 4,000 | 0.7513 | 3,005 |
| 4 | 585 | 0.6830 | 399 |
| | | | £5,602 |
| | | Less outlay | £5,000 |
| | | NPV | £602 |

**Investment B:**

| | £ | PV £1 at 10% | Discounted sum |
|---|---|---|---|
| Year | | | |
| 1 | 3,342 | 0.9091 | 3,120 |
| 2 | 3,342 | 0.8264 | 2,836 |
| 3 | 3,342 | 0.7513 | 2,578 |
| 4 | 3,342 | 0.6830 | 2,344 |
| | | | £10,878 |
| | | Less outlay | £10,000 |
| | | NPV | £878 |

On the basis of the NPV analysis, the investor should choose investment B which has the highest NPV.

(b) IRR analysis: It is clear from the NPV analysis that each investment has an IRR exceeding 10% because the NPV is positive.

**Investment A:**
**Trying 15%**

| | £ | PV £1 at 15% | Discounted sum |
|---|---|---|---|
| Year | | | |
| 1 | 600 | 0.8696 | 522 |
| 2 | 2,000 | 0.7561 | 1,512 |
| 3 | 4,000 | 0.6575 | 2,632 |
| 4 | 585 | 0.5718 | 334 |
| | | | £5,000 |
| | | Less outlay | £5,000 |
| | | NPV | £0,000 |

**Investment B:**
**Trying 14%**

| | £ | PV £1 at 14% | Discounted sum |
|---|---|---|---|
| Year | | | |
| 1 | 3,432 | 0.8772 | 3,011 |
| 2 | 3,432 | 0.7695 | 2,639 |
| 3 | 3,432 | 0.6750 | 2,317 |
| 4 | 3,432 | 0.5921 | 2,033 |
| | | | £10,000 |
| | | Less outlay | £10,000 |
| | | NPV | £0,000 |

On the basis of the IRR analysis, the investor should choose investment A, because it has the highest IRR.

In this example, the IRR and NPV methods give different rankings.

## Which investment should be chosen?

In investment B, an extra £5,000 has been employed. This can produce certain extra benefits. To analyse the situation further it is necessary to tabulate the difference in cash flows between investment A and B.

| | A | B | B-A |
|---|---|---|---|
| Outlay | £5,000 | £10,000 | £5,000 |
| Receipts: | | | |
| Year 1 | £600 | £3,432 | £2,832 |
| Year 2 | £2,000 | £3,432 | £1,432 |
| Year 3 | £4,000 | £3,432 | −£568 |
| Year 4 | £584 | £3,432 | £2,848 |

## Incremental analysis

The final column could be called the increment of B over A. This in itself becomes a cash flow on which an NPV or IRR analysis could be carried out. The cost of capital is 10%, and, as it has been assumed to represent the investor's target rate, it is also assumed that this is a rate of return that could be earned elsewhere. If the increment earns a return of less than 10%, the investor would be wise to invest £5,000 in project A and obtain 10% interest elsewhere with the remaining £5,000.

Another way of looking at this is to remember that the cost of capital could also represent the cost of borrowing money. If the investor's £10,000 has been borrowed, a return of less than 10% on the increment of £5,000 of project B over project A means that the loan charges on this £5,000 are not being covered and the loan of the second £5,000 was not worthwhile. However, if the increment can be shown to produce a return in excess of 10%, the loan charges will be covered. Investment B thus uses the whole £10,000 to good effect. If project A were chosen, the extra £5,000 could only earn 10%, as it is assumed that no return in excess of the cost of capital could be earned without incurring an extra element of risk.

This is called incremental analysis and is shown below for the increments of B over A of £2,832; £1,432; -£568; and £2,848.

| | £ | PV £1 at 10% | Discounted sum |
|---|---|---|---|
| Year | | | |
| 1 | 2,832 | 0.9091 | 2,575 |
| 2 | 1,432 | 0.8264 | 1,183 |
| 3 | - 568 | 0.7513 | −427 |
| 4 | 2,848 | 0.6830 | 1,945 |
| | | | £5,276 |
| | | | −£5,000 |
| | | NPV[1] | £276 |

[1] Any project incorporating frequent sign changes can produce more than one IRR due to the polynomial nature of the underlying equation.

The IRR of the increment exceeds 10%, so investment B should be chosen. The first £5,000 employed is as profitable as it would be if used in investment A; the second £5,000 is used more profitably than is possible elsewhere. B is therefore preferable to the investment of £5,000 at the cost of capital rate of interest of 10%.

# Summary

This chapter has demonstrated the use of the PV function to assess the acceptability of investment opportunities measured against an investors target rate of return. In those cases where the results of an NPV and an IRR analysis conflict, an incremental analysis may be undertaken to assess the return on any additional capital used.

The discounted cash flow process has the following key components:

- A NPV discounts the net benefits of an investment using the present value function at the investor's target rate or opportunity cost of capital rate. The initial outlay is deducted from the NPV and if the resultant figure is positive then the investment is acceptable at the target rate.

- An IRR which represents the investment return and is calculated by trial and error using an iterative process; this is facilitated with programmes on financial calculators, IRR and Goal Seek functions in Excel. Generally, the higher the IRR the better the investment opportunity given the same risk assessment.

- Care needs to be taken in assessing the benefits and their timing accurately and the analysis needs to include any resale or terminal value or, in some cases such as mining reclamation, a terminal cost.

- Cash flows with negative sums arising during the life of an investment may produce multiple values of IRR, hence careful and logical thought should be used to identify the IRR most likely to be correct.

# Spreadsheet user

Spreadsheets are ideally suited to financial modelling, the application of discounted cash flow solutions and the calculation of IRR. Indeed, most spreadsheets have financial functions such as NPV and IRR built in to save time and trouble. However, as explained below, the use of the built-in functions must be undertaken carefully as they may not operate in the way in which the user expects.

## Project 1: Calculating the NPV

You are evaluating a scheme which will cost £80,000 immediately and generates the following cash flow at the end of each year:

| Year 1 | £10,000 |
| Year 2 | £25,000 |
| Year 3 | £35,000 |
| Year 4 | £30,000 |
| Year 5 | £20,000 |

## Calculate the NPV of the scheme and the IRR

Construct a spreadsheet which calculates the NPV. Use the Present Value of £1 formula in cell E5 designed to pick up the discount rate (*i*) from the amount entered by the user in cell 3 and copy this formula to cells E6:E9, i.e. in cell E5 should be the formula = $1/((1 + (\$F\$3))\wedge C5)$. In cell F5 enter the formula D5*E5 and copy to cells F6:F9.

To calculate the NPV in cell F10 use the formula = sum(F4:F9).

(The use of the built-in NPV and IRR functions are explored on the next page.)

**An example is shown below:**

| | A | B | C | D | E | F |
|---|---|---|---|---|---|---|
| 1 | | | | | P.V £1at | Discounted |
| 2 | | | YEAR | | Discount Rate =10% | Sum |
| 3 | | | | | | |
| 4 | | Expenditure | | -80000 | 1.0000 | -80000 |
| 5 | | Income | 1 | 10000 | 0.9091 | 9091 |
| 6 | | | 2 | 25000 | 0.8264 | 20661 |
| 7 | | | 3 | 35000 | 0.7513 | 26296 |
| 8 | | | 4 | 30000 | 0.6830 | 20490 |
| 9 | | | 5 | 20000 | 0.6209 | 12418 |
| 10 | | | NET PRESENT VALUE = | | | 8957 |
| 11 | | | | | | |

Note: The above assumes that the cell E2 are in a percentage format (see spreadsheet user 1); if not the values in the above formula require division by 100.

Your NPV calculator spreadsheet should now look like the screenshot below:

| | A | B | C | D | E | F | G | H |
|---|---|---|---|---|---|---|---|---|
| 1 | | | | | | | | |
| 2 | | | Year | | P.V.£1@ | P.V.£1@ | | |
| 3 | | | | | Discount Rate of | 10.00% | | |
| 4 | | **Expenditure** | | -80000 | | -80000 | | |
| 5 | | Income | 1 | 10000 | | 0.9091 | 9091 | |
| 6 | | | 2 | 25000 | | 0.8264 | 20661 | |
| 7 | | | 3 | 35000 | | 0.7513 | 26296 | |
| 8 | | | 4 | 30000 | | 0.6830 | 20490 | |
| 9 | | | 5 | 20000 | | 0.6209 | 12418 | |
| 10 | | | | | NET PRESENT VALUE = | 8957 | | |
| 11 | | | | | | | | |

Microsoft Excel - Spreadsheets.xls

File  Edit  View  Insert  Format  Tools  Data  Window  Help

Arial  10  B I U  % , 100%

The present value can be calculated directly from the cash flow and target rate using the NPV function.

Note: in both NPV functions it is assumed that the first cash flow occurs at the end of the first period, and subsequent cash flows at the end of each subsequent period. In many property and other examples, in reality the outlay will occur at the beginning of the first period as in this project example.

Try calculating the NPV in Project 1 using the NPV function in a spare cell expressed as:

*In Excel:* =NPV (0.1,D4:D9)

The figure produced is £8,143: which differs from our calculation using the present value formula.

This undervaluation is caused because the NPV function is treating the outlay of £80,000 as occurring at the end of Period 0, not at the beginning.

To use the NPV function in Excel it needs to be modified if, as in most cases, the initial outlay occurs at the beginning of period 0.

The initial cash flow must be isolated so the NPV function in Project 1 must be modified to:

= −80000 + NPV (0.1,D5:D9)

The result should be the same as the original calculation using the present value formula.

## Project 2: Calculating the IRR

The IRR function in both Lotus and Excel uses the iteration technique to calculate the discount rate at which the NPV for a given cash flow is equal to 0:

The IRR function in Excel: = IRR(values, guess) the internal rate of return of a series of cash flows (range), the guess rate being used only to start the iteration process.

To calculate the IRR in Project 1 enter the following in a spare cell = IRR (D4:D9,0.1).

Note: for the IRR function to work there must usually be at least one positive and one negative value in the cash flow. Problems may occur where the cash flow is non-standard and it changes sign several times.

Add the IRR to your spreadsheet in Project 1:

| | A | B | C | D | E | F |
|---|---|---|---|---|---|---|
| 1 | | | | | P.V £1at 10% | |
| 2 | | | YEAR | | | Discounted |
| 3 | | | | | | Sum |
| 4 | | Expenditure | | -80000 | 1.0000 | -80000 |
| 5 | | Income | 1 | 10000 | 0.9091 | 9091 |
| 6 | | | 2 | 25000 | 0.8264 | 20661 |
| 7 | | | 3 | 35000 | 0.7513 | 26296 |
| 8 | | | 4 | 30000 | 0.6830 | 20490 |
| 9 | | | 5 | 20000 | 0.6209 | 12418 |
| 10 | | | NET PRESENT VALUE = | | | 8957 |
| 11 | | | | | | |
| 12 | | | | | | 13.90% |

To do this, add the following formula to cell G12: = IRR(D4:D9,F3). Your NPV and IRR calculator spreadsheet should now look like this:

## Project 3: Goal Seek

Using your spreadsheet from Project 1, calculate the IRR using Goal Seek.

## Questions: Chapters 2 and 3

1: A sum of £250 is invested at 9.5% for 15 years. What sum will there be at the end of 15 years?

2: A capital sum of £1,500 is needed in 10 years. How much must be invested today assuming a compound rate of 8.5%?

3: A mortgage of £40,000 is arranged at 8%. What is the annual repayment if the term is for 30 years?

4: A sum of £1,000 is invested at the end of each year and earns interest at 16%. How much will this accumulate to in 10 years' time?

5: How much must be invested each year at 6% compound interest to produce £2,500 in 16 years' time?

6: You have expectations of becoming a partner in private practice in 10 years' time. This will cost you £25,000:

   (a) How could you provide for this?
   (b) How much would it cost at 12%?
   (c) How much would it cost today saving at 12%, if inflation is increasing the cost at 12%?

7: If you save £200 a year for 10 years in a building society at 7%:

   (a) How much will you have at the end of 10 years?
   (b) If you added a further sum of £2,000 at the beginning of year 5, how much would you have at the end of 10 years?
   (c) If you do (a) and (b) and leave it to accumulate for a further five years, how much will you have?

8: A shop is let at £10,000 a year for five years after which the rent will rise to £50,000 in perpetuity. What is its value on an 8% basis?

9: 8% Treasury Stock are selling at £80 per £100 face value certificate. What is the true rate of return (IRR Gross Redemption Yield) if the stock has exactly five years to run to redemption? Try this manually with present values and using the IRR linear equation; and by using Goal Seek.

# Chapter 4

## Basic Principles

## Introduction

Valuation by comparing the subject property directly with prices achieved on the sale of comparable property in the market is the preferred method of valuation for most saleable goods and services. Valuation by this method is reliable provided that the sample of comparable sales is of sufficient size to draw realistic conclusions as to market conditions. This requires full knowledge of each transaction. Such a situation rarely exists in the market for investment property and, in the absence of directly comparable sales figures, the valuer turns to the investment method. The investment method is used for valuing income-producing property whether freehold or leasehold, because as a method it most closely reflects the behaviour of the various parties operating in the property market.

Valuation was earlier summarised as the estimation of the future benefits to be enjoyed from the ownership of a freehold or a leasehold interest in land or property, expressing those future benefits in terms of present worth. The valuer must be able to assess these future net benefits *(income)* and be able to select the appropriate rate of interest or yield to discount these benefits to their present value. However, assessment of the income pattern and the rate(s) of interest to be used to discount that income flow requires the valuer to complete a wide range of investigations and enquiries.

## Definitions

In a discipline that is derived from urban economics and investment analysis, one would expect to find some common agreement as to the meaning of terms used by practitioners. This does not exist, so that additional problems may arise when advice is given to investors who are more acquainted with terms used by other financial advisors. The advice of the most expert valuer is of minimal value if it is misinterpreted, so a definition of the terms used from hereon may be useful.

The explanations and definitions that follow may not achieve universal acceptance but are adhered to within this text and, where indicated, are defined by the Royal Institution of Chartered Surveyors (RICS) and Institute of Rating and Revenue Valuers (IRRV). The RICS *Valuation Standards* (The Red Book) (2011) contains a glossary of terms.

The question of value definition has bemused valuers and philosophers for centuries. It is recorded that Plato described 'the notion of value as the most difficult question of all sciences'. (Cited in *Real Estate Appraising in Canada* (The Appraisal Institute of Canada) (1987).)

An early definition is set out in the 1938 edition of Mustoe, Eve and Ansteys *Complete Valuation Practice*:

> In ordinary speech the 'value' of a thing means the amount of money which that thing is worth in the open market. The valuer, whether he is valuing real property or chattels or livestock, endeavours to assess each article in terms of pounds, shillings and pence; to him 'value' means the amount of money for which property will exchange.

Considerable difficulties emerged in the 1970s and again in the 1990s, over the meaning and definition of value. The RICS addressed the issues through the Asset Valuation Standards Committee in the 1970s, and reviewed the issue through a working party set up in 1993 under Michael Mallinson. Many of the recommendations of the Mallinson report have been accepted by the RICS, which now requires valuers to use the International Valuation Standards (IVS) definition of market value (MV). In the RICS definitions the italics are those used by the RICS and may be found defined in the RICS glossary.

## Market value

This definition is the one adopted by the RICS; it is an international definition and was developed by the International Valuation Standards Committee (IVSC):

> The estimated amount for which an asset or liability should exchange on the *valuation date* between a willing buyer and a willing seller in an arm's-length transaction after proper marketing wherein the parties had each acted knowledgeably, prudently and without compulsion.
> **(RICS VS 3.2.).**

The definition is set out in VS 3.2, together with a Conceptual Framework as published in International Valuation Standard 1; this conceptual framework explains or defines all the terms used in the definition.

## Value in exchange

This is a term used by economists and others and has the same meaning as MV. If a good or service is incapable of being exchanged for other goods or services or money equivalent, then it has no MV.

## Investment value or worth

This is defined in the Glossary of Terms in The Red Book as:

> The value of an asset to the owner or a prospective owner. (May also be known as worth.) (RICS VS Glossary).

Investment value is a subjective assessment of value to an owner and may reflect an owner's, or prospective owner's, perception of the market rather than a valuer's objective analysis of the market. The extent to which investment value differs to MV should be set out clearly by the valuer and it should not be described as a valuation.

## Price

Price is a historic fact except when qualified in such a phrase as 'offered at an asking price of …'. Under perfect market conditions value in use would equate with value in exchange and price would be synonymous with value. The property market is not perfect, and price and MV cannot always be said to be equal.

## Valuation

This is the art or science of estimating the value of interests in property. According to many dictionaries, the word 'valuation' is interchangeable with the word 'appraisal'. The latter word is used by Baum and Crosby (2008) as a source term covering both the assessment of MV and for the estimation of worth. The latter is better considered as analysis. In the market the term 'appraisal' is sometimes used in reference to the 'Appraisal of a Development Scheme' or as implying a process that is more comprehensive than a mere opinion of value. For clarity, this text uses the terms valuation and analysis.

Valuation used to be defined by the RICS (VS 2010) as:

A valuer's opinion of the value of a specified interest or interests in a *property*, at the *date of valuation*, given *in writing*. Unless limitations are agreed in the *terms of engagement* this will be provided after an inspection, and any further investigations and enquiries that are appropriate, having regard to the nature of the *property* and the purpose of the *valuation*.

## Valuation report

This is the formal presentation of the valuer's opinion in written form: 'The report must clearly and accurately set out the conclusions of the valuation in a manner that is not ambiguous or misleading, and does not create a false impression' (VS 6.1). As a minimum it must contain a sufficient description to identify the property without doubt; a value definition; a statement as to the interest being valued and any legal encumbrances; the effective date of the valuation; any special feature of the property or the market that the valuer has taken special note of; and the value estimate itself. (See Appendix B, and VS6.)

These definitions may be confusing to the reader and not obviously that important, but in practice they may be very important. Even more important is the valuer's instruction and the importance of communication between valuer and client: 'How much is it worth?'; 'What price should I offer?'; 'Is it worth £x million?'; 'What figure should we include in our accounts for the value of our property assets?'; and 'Is that the same as their market worth?', are questions which may give rise to different responses from the valuer and might result in the valuer expressing different opinions for the same property interest for different purposes. The basis or definition must be clarified at the time of agreeing instructions and must be confirmed in the valuation report.

In terms of this book and the methods set out herein, the terms 'value' or 'valuation' refer to the assessment of MV unless otherwise stated.

## The income approach

The income approach or investment method is a method of estimating the present value of the rights to future benefits to be derived from the ownership of a specific interest in a specific property under given market conditions. In property valuation these future rights can usually be expressed as future income (rent) and/or future reversionary capital value. The latter is in itself an expression of resale rights to future capital benefits. The process of converting future income flows to present value capital sums is known as capitalisation, which in essence is the summation of the future benefits, each discounted to the present at an appropriate market-derived discount rate. The terms *discount rate* and *capitalisation rate* are increasingly preferred to 'rate of interest', 'interest rate' or 'rate of return'. The use of the term 'rate of interest' should be restricted to borrowing, it being the rate of interest charged on borrowed funds or the rate of interest to be earned in a savings account.

A distinction needs to be made between return on capital and return of capital. The interest rate or rate of return refers to return on capital only and is sometimes referred to as the remunerative rate to distinguish it from the return of capital or sinking fund rate. Valuers, when capitalising income, are technically discounting future benefits but use a capitalisation rate or all risks yield; these yields reflect not only market risks but also any positive benefits in the form of income growth. Valuers undertaking a valuation or assessment of worth use discount rates; these do not reflect any expectation of income growth as this is included in the estimates of future income.

It is necessary to discount future sums at a rate to overcome:

- Liquidity: that is the time and costs associated with a disposal of property together with sale price uncertainty. Sales of investment property due to their heterogeneous nature rarely sell at their valuation, RICS' reports on this matter suggest that only 80% sell within ± 10%; sale costs will be in the region of 0.5% to 1.0%

- Time preference: the risks associated with uncertainty about the future and the need to allow for the fact that money in the future is likely to have less purchasing power than the same money today or, expressed differently, money can be saved to earn interest today and therefore an investment today for future money benefits must deliver a similar return plus an added amount for the additional risks associated with property.

Some of these risks have been identified by Baum and Crosby in their authoritative text on Investment Valuation and include:

- TENANT RISK
  - Voids
  - Non payment of rent
  - Breach of other covenants.
- SECTOR RISK
  - Type specific, e.g. retail, offices
  - Location specific.
- STRUCTURAL RISK
  - Building failure
  - Accelerated depreciation or obsolescence.
- LEGISLATION RISK
  - New laws specific to property that might affect usability, rents, etc.
- TAXATION RISK
  - Possibility of future taxes on property.
- PLANNING RISK
  - Planning policy may impinge upon performance, e.g. new roads, pedestrianisation of roads.
- LEGAL RISK
  - Undiscovered issues affecting legal title, e.g. rights of way.

The uncertainty associated with the macro economy and inflation must also be added to this list. A capitalisation rate reflects all these factors and more and is referred to as an all risks yield (ARY or cap rate).

In its simplest form income capitalisation is merely the division of income ($I$) by the capitalisation rate ($R$) (*which is expressed by valuers as a decimal i*).

A rate of capitalisation is expressed as a percentage, say ˇ10%, but as was stated in Chapter 1, all financial analyses and calculations require this ratio to be expressed as a decimal. Therefore: 10% becomes 0.10, 8% = 0.08, 6% =0.06, etc., and thus Value ($V$) = $I/i$.

---

**Example 4.1**

Calculate the present value of a freehold interest in a property expected to produce a rent of £10,000 pa in perpetuity at 10%.

$$V = I/i = \frac{10,000}{0.10} = £\,100,000$$

In preference to dividing by $i$, UK valuers have always used the reciprocal of $i$, and multiplied the Income I by a capitalisation factor which is called the Years' Purchase or YP.

Thus:

$$V = I \times (1/i) = £\,10,000 \times \frac{1}{0.10} = £\,10,000 \times 10 = £\,100,000$$

The capitalisation of an infinite income, dividing by a market rate of capitalisation, is a simple exercise known as capitalisation in perpetuity. For many valuations this is all that is necessary and some valuers customarily create very simple tables to undertake investment valuations as closely as possible to a form of direct capital comparison. This is possible in defined markets where the yield variation will be in a narrow range, and the market rent for that type of property in that locality is also known to be in a narrow range. This process is useful to establish an approximate level of value before undertaking the more detailed analysis and valuations described later in this book.

For example, office space in the central business area of a town may typically let at net rents of between £150 and £175 per square metre, and freehold investment sales may consistently occur at yields of between 8% and 9%. The combinations of value per square metre can conveniently be presented using an Excel spreadsheet or set out to the nearest £ in a table.

| Rent/yield | 150 | 155 | 160 | 165 | 170 | 175 |
|---|---|---|---|---|---|---|
| 8% | 1875 | 1937 | 2000 | 2062 | 2125 | 2187 |
| 8.25% | 1818 | 1879 | 1939 | 2000 | 2060 | 2121 |
| 8.5% | 1765 | 1823 | 1882 | 1941 | 2000 | 2059 |
| 8.75% | 1714 | 1771 | 1828 | 1886 | 1943 | 2000 |
| 9% | 1667 | 1722 | 1778 | 1833 | 1889 | 1944 |

In a relatively stable market such a table may be useable for a year or more. It also provides a useful comparative check against other properties and sales that have occurred. The valuation of a 1,500 square metre office building now requires the valuer to assess, based on comparison with other known lettings and sales, the rent and yield – say £160 and 8.75% and the valuation becomes $1,500\,m^2 \times £1828 = £2,742,000$, which the experienced valuer might round to £2,750,000. The trainee valuer may well be advised to become aware of ball park figures such as these, as they provide a useful check when undertaking more complex investment valuations. If the property being valued, for example, was currently let at £140 per $m^2$ with a rent review in two years' time to a potential market rent of £160 per $m^2$ then the value of the encumbered freehold (that is the freehold subject to the current lease arrangements) must be less than £2,750,000. In a similar vein, in some markets, commercial properties, such as business park units, sell at a market price per $m^2$ and valuation is possible by using direct capital comparison.

Where, however, the income is finite, an allowance must be made for recovery of capital. Two concepts exist in UK valuation practice: the internal rate of return (IRR) and the sinking fund return. Each allows for the systematic return of capital over the life of the investment, but each approach is based on different assumptions.

The IRR assumes that capital recovery is at the same rate as the return on capital $i$. It reflects the normal investment criterion of a return on capital outstanding from year to year and at risk, with the capital being returned from year to year out of income. Where capital recovery is at the same rate as the risk rate, the present value of £1 per annum (YP single rate) is used which in turn is the recipro-cal of the annuity £1 will purchase. Thus, any finite income stream can be treated as an annuity calculation. This concept is the more acceptable because the present worth of any future sum is that sum which if invested today would accumulate at compound interest to that future sum, and hence the present value of a number of such sums is the sum of their present values. This is sometimes referred to as The Inwood Premise, namely that an income flow has a present value based on a given discount rate.

If the sums are equal and receivable in arrears then:

$$V = 1 \times \left( \frac{1 - PV}{i} \right)$$

If the sums vary from year to year then:

$$V = \sum \frac{I_1}{(1+i)} + \frac{I_2}{(1+i)^2} + \frac{I_3}{(1+i)^3} \cdots \frac{I^n}{(1+i)^n}$$

The sinking fund return assumes capital is returned by a fixed annual reinvestment in a separate fund, possibly accumulating at a different rate of interest, which is sometimes known as the Hoskold principle.

Mathematically, there need be no distinction between the two concepts, as capital recovery can always be provided within a capi-talisation factor by incorporating a sinking fund to recover capital:

$$PV £ 1pa = \frac{1}{i + ASF}$$

### Example 4.2

Calculate the value today of the right to receive an income of £1,000 at the end of each year for the next five years if a return of 10% is to be received from the purchase.

| Income | £1,000 |
|---|---|
| PV of £1 pa for 5 years at 10% | |
| (YP) | 3.7908 |
| | £3,790.80 |

Proof IRR Basis:

| Year | Capital | Income | Interest at 10% | Capital recovered |
|---|---|---|---|---|
| 1 | 3,790.80 | 1,000 | 379.80 | 620.92 |
| 2 | 3,169.88 | 1,000 | 316.99 | 683.01 |
| 3 | 2,486.87 | 1,000 | 248.68 | 751.32 |
| 4 | 1,735.55 | 1,000 | 173.56 | 826.44 |
| 5 | 909.11 | 1,000 | 90.91 | 909.09 |

Proof Sinking Fund (SF) Basis

| | |
|---|---|
| Capital | £3,790.80 |
| Annual sinking fund to replace £1 at 10% over 5 years = | 0.16380 |
| | £620.9333 |

Therefore:

$$\text{Income net of capital recovery} = £1,000 - £620.933$$
$$= £379.067 \text{ and}$$
$$= (£379.067/£3790.8) \times 100 = 10\%$$

The most widely sold time-restricted investments are government gilts which are bought and sold on the basis of their gross redemption yield which is another term for IRR. No buyers of government gilts look to recover capital by any form of sinking fund reinvestment. For this reason the IRR concept is the one that should be accepted for all time-restricted investments. However, if it can be shown that investors do require a return on initial outlay throughout the life of the investment, then the sinking fund concept can be adopted. Further, if it can be shown that investors expect a return of capital at a different rate to the return on capital then this can be allowed for in the formula (see page 22). It should be noted that the shorter the life of an investment, the greater will be the affect of tax for the taxpaying investor who needs to be aware of the before and after tax rates of return. However, on this point, Charles B Akerson, writing in 1973 in the American Institute of Real Estate Appraisers *Capitalization Theory and Techniques Study Guide*, noted that

> the Hoskold premise differs from the Inwood premise in that it utilizes two separate interests rates 1) a 'speculative rate' representing a fair rate of return on capital commensurate with the risks involved, and 2) a 'safe rate' for a sinking fund designed to return all capital in a lump sum at the termination of the investment..'

He further noted that 'The system has merit, but now has limited acceptance'.

Here and elsewhere in the book, the authors entirely agree with this comment and strongly argue against the use of this dual rate or sinking fund return concept.

The arithmetic manipulation of figures in capitalisation exercises is not, of course, valuation. Valuation is a process which requires consideration of a number of variables before figures can be substituted in mathematically proven formulae. Any assessment of present worth or MV can only be as good as the data input allows and that factor is dependent upon the education, skill and market experience of the valuer. The ability to analyse and understand the market is of paramount importance. On this point and in the context of the discussion on IRR and sinking fund rate of return it must again be noted that the only true analysis that can be made of an investment sale price is the establishment of the IRR. The relationship between a price and a known finite income can only be analysed to assess the sinking fund return if a prior decision is made as to the rate of interest at which the sinking fund is to accumulate.

Valuation has been likened to a science, not because of any precision that may or may not exist, or because in part it involves certain basic mathematics, but because the question *'How much?'* poses a problem that requires a solution.

The scientific approach to problem solving is to follow a systematic process. There may well be short cuts within the process. Indeed, the discounting exercise itself, because it is repetitive, can frequently be carried out by pre-programmed calculators and computers. Short cuts exist if data is already available, but adoption of a systematic approach provides the confidence that full account has been taken of all the factors likely to affect the value of a property. This systematic approach is outlined in the next section entitled 'Valuation process'.

## Valuation process

First, the valuer should define the problem by asking the following questions: 'What are the client's real requirements?'; 'Why does he or she want a valuation?'; 'What is the purpose of the valuation?'

These questions should establish whether the client requires a market valuation or a valuation for company asset purposes, insurance or rating.

Following the Mallinson report, RICS has reconfirmed the need for the valuer to clarify and agree terms of engagement; that is, to confirm the client's instructions and to do so in writing. This stage in the process is a mandatory requirement for RICS and IRRV members, the details are set out in the RICS *Valuation Standards* (2011). Most students can readily access this through their libraries or learning centres online. An additional resource for most surveying students is the online RICS Books isurv valuation service (www.isurv.co.uk). RICS valuers are required, in brief, to: identify the client; the purpose; the subject matter; the interest; the type of property and how it is used; the basis or bases of value; the valuation date; disclosure of involvement statement; status of the valuer; currency; assumptions and special assumptions; the investigations to be made; the information to be relied on; the need for consent prior to publishing a valuation; limitations on third party liabilities; confirmation that the valuation is in accordance with the RICS standards; basis on which the fee is to be calculated; complaints handling procedures; and a statement that the valuation may be investigated by the RICS for regulatory purposes (see RICS Valuation Standards VS1 and UKVS4 Regulated Purpose Valuations). Whilst much of this may seem to be obvious, most disagreements relate to misunderstandings arising from poor terms of engagement; and it has been known for the wrong property to be valued so it is important to begin with a clear agreement of 'what', 'when' and 'how' and everything else.

Having established the fact that the instructions are within the valuer's competence, and having confirmed the instructions and the date of the valuation, the valuer needs to inspect the property, collect all the data needed for the valuation and carry out all appropriate investigations.

Within specific economic and market conditions, the valuer will be considering five principal qualities:

- Quality of the legal title.
- Quality of the lease(s).
- Quality of the location.
- Quality of the building.
- Quality of the tenant(s) in occupation or those that can be assumed in the case of a new or owner-occupied property.

### Quality of the legal title

In the majority of cases, the title will be freehold or leasehold but commonhold is possible. Checks will be made for any restrictive covenants which might restrict the use of the property or the density of development; rights of way, rights of light and where let all leases will be read.

### Quality of the lease(s)

Investors prefer long leases of 20 to 25 years, with regular rent reviews and that have a medium to long unexpired term. The terms and conditions in respect of repairs, insurance and management should be such as to leave the rent payable as a net income to the landlord. For an individual building, a lease on full repairing and insuring terms is preferred. In multi-tenanted properties, a service and maintenance charge should be included to provide full recovery of all expenses including the associated costs of management. The rent review should be upward only, index linked or stepped. The user should not be overly restricted as this can impact on market rents. Assignment or subletting should only be with the landlord's consent. Alterations and improvements should also only be with the landlord's consent.

### Quality of the location

Similar properties can have very different values, due simply to location. Environment checks will be made to check for flood risks and water runoff, contamination risks, crime levels, access, parking and any other factor which might impact on the suitability of the location for the permitted use (see RICS, *Contamination, the Environment and Sustainability; Implications for Chartered Surveyors and their Clients*, 3rd edition (2010) for details of matters to be checked during an inspection and online).

### Quality of the building

The valuer must be able to assess the quality of the building in terms of its construction and current suitability for tenants seeking buildings for that permitted use. New buildings are likely, due to recent changes to building regulations and carbon footprint requirements, to be built to a higher occupation standard and to be greener. Older buildings may have poor energy ratings, may be costly to maintain and may not be suitable for today's market due to floor to ceiling heights or room areas. Space requirements change over time, particularly for retailers; typically for many there are no longer store rooms and the space needs to be sufficient for all stock to be on display.

## Quality of the tenant(s)

Where tenanted, the valuer will be checking the tenant's covenant strength or credit rating. It is important to judge whether the tenant(s) are capable of paying the rent and of meeting their other obligations under the lease. Checks can be made through Dunn and Bradstreet, Experian, banks and suppliers. Other checks will include Company House checks on annual reports and accounts and, in the case of quoted companies, share value and share movement. Tenants can be plcs, llcs, partnerships, individuals, government departments and charities; and within each sector and subsector there will be tenants deemed to be of sound covenant and others where there will be added risk of non-payment. The 2007 credit crunch has been a strong warning not to assume anything about a tenant without very careful analysis.

The factors which make a building and location good for a tenant will also make it attractive to an investor.

RICS VS 5 specifies a minimum of matters to be considered by a valuer in the process of the valuer's investigation. These, together with factors listed below, provide some idea of the mass of information that needs to be collected and assessed in order to arrive at an objective opinion of the property's marketable qualities, which in turn colour the valuer's judgment as to the income-generating capabilities of the property and its suitability as an investment.

RICS Valuation Standards VS 5.1.5 contains the following guidance:

Many matters which become apparent during the *inspection* may have an impact on the market's perception of the value of the property. These can include:
a) the characteristics of the surrounding area, and the availability of communications and facilities which affect value;
b) the characteristics of the property;
c) the dimensions, and areas, of the land and buildings;
d) the construction of any buildings and their approximate age;
e) the uses of the land and buildings;
f) the description of the accommodation;
g) the description of installations, amenities and services;
h) the fixtures, fittings and improvements;
i) any plant & equipment which would normally form an integral part of the building;
j) the apparent state of repair and condition;

k) environmental factors, such as:

abnormal ground conditions, historic mining or quarrying, coastal erosion, flood risks, proximity of high-voltage electrical equipment;

l) contamination, such as:

potentially hazardous or harmful substances in the ground or structures on it, for example, heavy metals, oils, solvents, poisons or pollutants that have been absorbed or integrated into the property and cannot be readily removed without invasive or special treatment, such as excavation to remove subsoil contaminated by a leaking underground tank or the presence of radon gas;

m) hazardous materials, such as:

potentially harmful material in a building or on land but which has not contaminated either. Such hazardous materials can be readily removed if the appropriate precautions and regulations are observed, for example, the removal of fuel (gas) from an underground tank or the removal of asbestos;

n) deleterious materials, such as:

building materials that degrade with age, causing structural problems, for example high alumina cement, calcium chloride or woodwool shuttering;

o) any physical restrictions on further development, if appropriate.

Further detailed guidance for the UK is provided in UKGN4 (2011).

Measurements are taken in accordance with the current RICS *Code of Measuring Practice* (2007); this is currently the 6th edition. These will normally include measurements of all the land, the gross external area (GEA) of the buildings and either the net internal area (NIA) of the buildings or the gross internal area (GIA). The RICS definitions are as follows:

- 'GEA is the area of a building measured externally at each floor level.'
- 'GIA is the area of a building measured to the internal face of the perimeter walls at each floor level.'
- 'NIA is the usable area within a building measured to the internal face of the perimeter walls at each floor level.'

The critical difference between GIA and NIA is the word 'usable', this means that NIA excludes a wide range of factors which are included in GIA such as: common areas, toilet areas, lifts, stairwells, permanent circulation areas, structural walls, columns, etc. Valuers have to be familiar with the RICS codes and their application as it is essential for consistency within the market; the code specifies 22 different applications (APP1 to APP22) so it is both the definition with the stated inclusions and exclusions and their application that must be known and understood. For planning purposes GEA is normal. For valuation purposes business uses are measured on an NIA basis; NIA for shops and offices; GIA for department stores, food superstores, retail warehouses; GIA is normally used for warehouses and industrial buildings, but there are some exceptions to this when NIA will be used. In the case of shops for valuation and letting purposes, NIA may be called retail area RA.

Not all the answers come from the inspection, although good observation during an inspection will alert the valuer to many potential issues which will need to be reviewed in formulating an opinion of value. As part of the process of inspection and investigation, the valuer will undertake desk searches of historic uses, searches of the environmental agency's sites, and will check planning and highways matters with the relevant authorities. The actual use of the property may not always comply with current planning consents. Other checks include checks on title, full review of all the leases, credit and other financial checks on tenants. As part of the building inspection, notice will have been taken as to whether the building meets with the requirements of building regulations, health and safety, disability compliance. Note will also be taken as to whether all required certificates are current, such as fire certificates, energy certificates, lift inspection certification, etc.

In terms of the economy, the valuer will be considering the national and local economic picture. Looking at the economic base of the area, employment and unemployment data; wages; population structure; growth prospects; stock of similar property; vacancy rates; rent levels and investor interests, all will be considered within the context of the property market and the alternative competing investment markets.

This data is collected for analysis if considered significant in terms of value. The valuer needs to know what other properties are in the market, whether they are better in terms of location, etc., who is in the market as potential buyers and as potential tenants, and whether the market for the subject property is active.

To collect the level of information required entails either very sound local knowledge, or the need to make enquiries of the local planning authorities, rate collecting authorities, highway authorities, transport companies, local census statistics and a range of other sources, including issues such as crime rates.

## Factors affecting investor yield requirements and market capitalisation rates

The following list is not comprehensive but is indicative of those factors that a valuer must have regard to:

- General level of interest rates.
- State of the national and local economies.
- Government financial and fiscal policies.
- Location of the property in terms of labour, transport and general and special accessibility.
- Legal factors such as planning, highways, landlord and tenant legislation, health and safety, disability requirements, fire, working conditions, EU directives.
- Rental growth prospects.

- Capital growth prospects.
- Security of rental income, tenants covenant, unexpired lease term, rent reviews.
- Taxation of income and capital, VAT, Stamp Duty Land Tax (SDLT).
- Liquidity issues both cost of disposal and probable time to realise the value of the property.
- Volatility of the market.
- Management costs.

The major firms of surveyors, valuers and property investment organisations have electronic call on a wide range of data, which is available to the general public on the internet, and to and subscription services of special interest to the property market. Large practices and major investors in property have research departments which are continuously analysing the economy, sectors of the economy and regions. This information source is kept up-to-date by research members extracting relevant information from national and local papers, local council minutes, etc. There are now many online sources of economic, market and general data relevant to real estate, including access to data on property transactions held by the Land Registry. Due diligence means that for many valuers ignorance of such data and their failure to make appropriate use of any such relevant data could result in action by clients for negligence.

Having collected the information shown above, it is then necessary to consider the market.

> 'A major danger in assessing the direct property market lies in failing to identify the main segments into which the market is divided according to the value, reversionary terms, etc.… of a property.'

This comment was made in the Greenwell & Co Property Report in October 1976 but is as true today as it was 30 years ago.

The valuer must be able to identify:

- The most probable type of purchaser.
- The alternative comparable properties on the market for that purchaser market.
- The current level of demand within that sub market.
- Any new construction work in hand which will offer better/ newer and, today, more energy efficient space for tenants, which will impact on investor demand for existing properties.

These and other questions will indicate what market data is required as preparatory material for the valuation. A particular concern to valuers is the need to identify the possibility of the property containing deleterious materials and/or of the site or buildings being contaminated.

The most difficult aspect of an income capitalisation exercise is the determination of the correct capitalisation rate. Every property investment is different, and if the available data on sales are insufficiently comparable then it may become difficult to justify the use of a selected rate. Thus, if a property with a five-year rent review

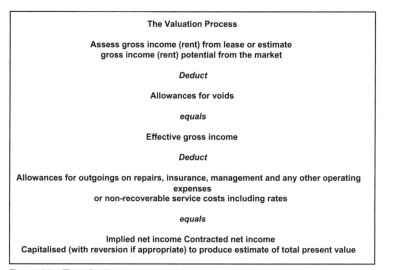

**Figure 4.1** The valuation process

pattern sells on a 7% basis, then this can be assumed to be the market capitalisation rate for that type of property let on a five-year review pattern, with the first review at a comparable date in the future and with a comparable level of rental increase.

If the subject property differs in any respect, then the valuer will need to seek better comparables or will have to adjust that rate to reflect the differences between the comparable property and the subject property. That adjustment will be based on an appreciation of the market impact of all the factors set out above. In principle the better the property, the better the location, the better the tenant, the better the location, and the better the lease, the lesser will be the risk and the lower will be the capitalisation rate and vice versa.

The valuation process can be redefined as shown in Figure 4.1. Within this process the three main variables likely to have the greatest effect on the final estimate of value can be identified: income (rent), operating expenses (landlord's outgoings) and capitalisation rate (all risks yield or equivalent yield). It is our presumption that a valuer practicing in a well-run valuation department will be able to provide reliable figures for these items, and should be able to support the figures used from available analysed data with soundly argued reasons for any adjustments made to reflect market changes and variations between the subject property and the supporting comparables.

## Income or rent

Capitalisation is the expression of future benefits in terms of their present value. Valuation therefore requires the valuer to consider the future: but current UK valuation practice reflects a distrust of making predictions. The convention has developed of using initial yields or all risks yields on rack rented property as the capitalisation rates to be applied to current estimates of future rental income, thereby building the market's forecast of future expectations into the capitalisation rate. The forecast is still made but the valuer has avoided any explicit statement. American appraisers consider this approach to be outmoded and argue that it should only be used if a level constant income flow is the most probable income pattern. In other circumstances they recommend the use of a Discounted Cash Flow Approach.

The RICS research report (Trott, 1980) on valuation methods recommended the use of growth explicit models for investment analysis and suggested that a greater use should be made of growth explicit discounted cash flow (DCF) models in the market valuation of property investments. This view is supported by Baum and Crosby (2008).

Whenever the investment method is to be used, whether in relation to tenant-occupied property or owner-occupied property, an initial essential step is the estimation of the current rental value of that property. If the property is already let this allows the valuer to consider objectively the nature of the present rent roll. Initially, the most important task must be the estimation of rental income.

If the property is owner-occupied then it is necessary to assess the imputed rental income. The assessment requires analysis of current rents being paid, rents being quoted and the vacancy rate of comparable properties.

This assessment should be carried out in accordance with the RICS definition of market rent (MR), set out in The Red Book as follows:

> 'The estimated amount for which a property, or space within a property, should lease(let) on the valuation date between a willing lessor and a willing lessee on appropriate lease terms, in an arm's-length transaction, after proper marketing wherein the parties had each acted knowledgeably, prudently and without compulsion'

Again, 'this definition must be applied in accordance with the conceptual framework of MV at VS 3.2 together with the ….supplementary commentary' (VS 3.3).

It will be seen later that this differs to the statutory definition in the Landlord & Tenant Act 1954 and will differ from many sets of assumptions set out in existing leases. In assessing the rental value of a property, the valuer needs to check the basis for such calculations before turning to the market for supporting evidence (see Figure 4.2).

Detailed analysis of the letting market must precede analysis of a specific letting. Thus, if it can be demonstrated within the market for office space that such space users will pay a higher rent for ground floor space than for space on a higher floor, then this may be reflected in the way in which a letting of a whole office building is analysed in terms of rent per square metre. If the rent of luxury

| Situation | Basis |
|---|---|
| Lease renewals of business premises under the Landlord and Tenant Act 1954. | Section 34 of the Landlord and Tenant Act 1954. |
| Rent reviews. | Rent review clause within the current lease. |
| Vacant property, owner-occupied property or reversion to a new letting. | MR as defined by RICS in Red Book. |
| Estimates of MR. | MR as defined by RICS in the Red Book. |

**Figure 4.2**   Rental basis

flats and apartments can be shown to have a closer correlation with the number of bedrooms than with floor area then analysis might be possible on a per bedroom basis.

If floor area is to be used then continuing analysis will indicate whether or not bids are influenced by, for example: the size, quality and location of entrances, common areas, (providing toilet, vending and catering facilities); the provision of air conditioning, car parking, security, accessibility (in some cases to facilitate 24/7 working practices), lifts; communal facilities, recreation, leisure, food, banks, cash points, first aid; design and layout, floor plate and structural grid layout, flexibility; other matters such as energy efficiency and increasingly the provision of health and wellbeing facilities, including gyms and relaxation areas.

Reliable estimates of current rents can only flow from analysis of rents actually being paid, and under no circumstances should a valuation be based on an estimate of rent derived from analysis of an investment sale where there is no current rack rent passing. This is because rent analysed in this manner depends upon the assumption of a capitalisation rate.

For example, if the only information available on a transaction is that a building of $500\,m^2$ has just been sold for £500,000 with vacant possession, the only possible analysis is of sale price per $m^2$.

In most cases, analyses of rent being paid should be on the basis of NIA for offices and retail or GIA for other commercial and industrial properties, see RICS *Code of Measuring Practice*, but agricultural land is often quoted on a total area basis inclusive of farm dwellings and buildings.

It is critical to know what should be included as part of NIA and GIA and the measurements themselves must be taken accurately.

There is a growing tendency for shop premises to be let at a fixed rental plus a percentage of turnover. The construction of turnover rent agreements is very complex. In most cases a base rent will be payable of around 70% to 90% of the market rent, plus a specified percentage of the turnover achieved; specifying the calculation of this and the amount to be collected per quarter needs to be carefully set out in the lease. There is a link between the base rent and the turnover percentage and clearly an agreement can vary from 100% based on turnover and 100% on MR and anything in between. Where a valuation is required of property let on such terms, full details of total rents actually collected, checked against audited accounts for a minimum of three years should be used as the basis for determining income cash flows for valuation purposes.

In each case the principal terms and conditions of the lease must be noted. The following key terms can affect the level of negotiated rent:

- Total length.
- Rent reviews and frequency.
- Payment frequency, e.g. monthly or quarterly in advance.
- Incentives.
- Responsibility and liability for repairs, insurance and management.
- Restrictions on use and opening hours.
- Alienation (the right to assign or sublet the whole or part).

Some of these points are considered in more depth in Chapter 8 on landlord and tenant valuations.

Due to the heterogeneous nature of property, it is customary to express rent in terms of a suitable unit of comparison, thus:

- Agricultural land　　　　rent per hectare
- Offices and factories　　rent per m² GIA
- Shops　　　　　　　　rent per m² overall or per m² ITZA (in terms of zone A)

Analysis of rents to arrive at a comparable rent per m² is a relatively simple exercise of dividing the total rent by the GIA or NIA as appropriate. In the case of shops, the property industry has developed the concept of zoning.

## The concept of zoning

Three alternative approaches have been developed for analysing shop rent:

- Overall analysis.
- Arithmetic zoning.
- Natural zoning.

### Overall analysis

The rent for the retail space is divided by the lettable rental space (NIA) to obtain an overall rent per m². This is a simple approach but is complicated by the practice of letting retail space together with space on upper floors used for storage, sales, rest rooms, offices or residential accommodation at a single rental figure. In these circumstances it is desirable to isolate the rent for the retail space.

Some valuers suggest there is a relationship between ground floor space and space on the upper floors – this will only be the case where the user is the same or ancillary (e.g. storage). Here, custom or thorough analysis will indicate the relationship, if any, between ground floor and upper floor rental values. Where the use is different, the rent of the upper floor should be assessed by comparison with similar space elsewhere, and deducted from the total rent before analysis. Thus, if the upper floors comprise flats, then the rent for these should be assessed by comparison within the residential market and then deducted from the total rent.

Overall analysis tends to be used for shops in small parades and for large space users. In the latter case it is reasonable to argue that tenants of such premises will pay a pro rata rent for every additional square metre up to a given maximum. The problem here is that what one retailer might consider being a desirable maximum could be excessive for all other retailers. Such a point should be reflected as a risk, accounted for by the valuer in the consideration of a capitalisation rate.

## Arithmetic zoning

This approach is preferred in many cases to an overall rent analysis because, in retailing, it is the space used to attract the customer into the premises that is considered to be the most valuable (namely the frontage to the street or mall). Again, the rent for the retail space should be isolated before analysis. Where arithmetic zoning is used it is custom and practice to use three zones, two with depths of 6.1 metres and a remainder and to halve back the rent, that is to argue that the rent a tenant will pay for a square metre in zone B will be half the amount they would be willing to pay for a square metre of space in zone A (the first 6.1 metres). How this works can be seen in Example 4.3.

### Example 4.3

Analyse the rent of £50,000 being paid for shop premises with a frontage of 6 m and a depth of 21 m.

Overall = £50,000 ÷ 126 (6 × 21) = £396.82 per m$^2$

Zoning: Assume zones of 6.1 m depth and £x per m$^2$ rent for zone A

| | | | |
|---|---|---|---|
| Then zone A = 6 × 6.1 | = 36.60 m$^2$ at £x | = 36.60x |
| zone B = 6 × 6.1 | = 36.60 m$^2$ at £0.5x | = 18.30x |
| zone C = 6 × 8.80 | = 52.80 m$^2$ at £0.25x | = 13.20x |
| | | | 68.10x |

$$50,000 \div 68.10 = 734.21$$

The rental value for zone A is therefore £734 per m$^2$ (ITZA).

Analysis of market rents will reveal a pattern which may reflect the footfall of shoppers. There is likely to be a high spot, with rents diminishing the further one moves away from the highest footfall.

The observant reader will realise that the space could be divided into any number of zones. Different regions and different retail centres display different trading patterns and whilst a common convention is to use two zones and a remainder with zone A and B of 6.1 m depth, other practices will be found. Custom and practice do not necessarily reflect market behaviour and adapting different zones and zone depths for analysis can produce considerable variations in opinion as to the ITZA rent for a given location. Currently, the

Valuation Office Agency use zone A rents for rateable value calculations, in so doing they converted 20 foot zones to the nearest cm and rounded to 6.1 m for zone A and zone B; 6.1 m has therefore become the industry standard.

Certainly in practice retailers do not see the premises divided into rigid zones. Every rental estimate must be looked at in the light of the current market and common sense – who would be the most probable tenant for premises 7 m wide by 100 m deep? Is there any user who operates in such space, or is the last 50 m waste or valueless space for most retailers in that locality?

Upper floor and basement sales space is included in the analysis. The relationship with zone A must be determined by considering the level of sales or density of shoppers to be found in the first floor and basement areas compared to the ground floor; this might be 1/6th or 1/8th of zone A.

Having established the MR in terms of zone A, the valuer can now assess the MR for the shop due to be valued.

### Example 4.4

Estimate the market rent for a shop in a high street location of a market town. Zone A rent on a 15 year full repairing and insuring (FRI) lease with five-year upward rent reviews is £500. The shop trades on the ground and first floor and measures 8 m × 24 m. The first floor has an retail area (RA) of 150 m². First floor retail space by comparison is 1/8th of zone A. Then this could be set out as:

$$48.80 + \frac{48.80}{2} + \frac{94.40}{4} + \frac{150}{8} = 48.4 + 24.4 + 23.6 + 18.75$$
$$P = 115.55 \times £500 = £57,775$$

Or as:

| | | |
|---|---|---|
| 8 × 6.1 | = 48.8 | 48.80x |
| 8 × 6.1 | = 48.8 × 0.5x | 24.40x |
| 8 × 11.8 | = 94.40 × 0.25x | 23.60x |
| 150 | = 150 × 0.125x | 18.75x |
| | | 115.55x |

As x = £500 the MR is 115.55 × £500 which is £57,775.

### Natural zoning

This method can only effectively be used to analyse rents within a shopping street or centre where information is available on a number of units, as it requires comparison between units. As previously explained, the rents for retail space must be isolated from the rent for the premises as a whole. It assumes that there is a base rent payable for a given area of space and that the extra rent paid for larger areas represents the additional rent for the extra floor area. It is difficult to use as an ongoing basis of analysis. This technique was originally detailed in Emeny and Wilks (1982). As it is not in common use and only of academic interest, the reader is referred to the 5th edition of this book for further information.

In terms of the rental analysis of retail space, warehouse accommodation and factory space, the valuer needs to have a sound knowledge of the specific requirements of different retailers and of different manufacturers.

Dealing with standard shop units, for example within a shopping mall, the unit of comparison can often be left as the 'shop unit'. Rent will be a factor of location/position and not size.

The developing technique, although not yet found to be any more satisfactory than those listed, is multi-variate analysis, when the dependent variable rent is considered against a number of independent variables which could include size, location, distance from car parks, bus station, etc.

It would be ideal if all rental estimates could be based on true comparables, i.e. those of the same size, design, facilities, location, etc. This is rarely the case in practice, however. With experience, adjustments can be made for some of these variations, but wherever possible, estimates of rent should be based on close comparables.

As far as office records are concerned, strict procedures should be adhered to so that the format of data is consistent. In this respect it is recommended that all rents are analysed net of landlord's outgoings (see below) and that apart from obvious factors such as the address of the property, the record should contain details of facilities included (e.g. central heating, air conditioning) and the lease terms.

The valuer must be in possession of all the facts of a given letting before any rental figure can be analysed. In the case of residential and commercial property, statute intervenes to protect the tenant in several ways (see Chapter 10). Thus, a tenant of business premises who carries out improvements would not expect to pay an increase in rent for these improvements on the renewal of the lease. The valuer must therefore know the basis upon which a given rent was agreed before he or she can make proper use of the information. An earlier lease might have been surrendered or the tenant might have undertaken to modernise the premises. Either of these could have resulted in a rental lower than a market rental being agreed between the parties. Of increasing concern is the lease definition of rent where there is provision for a rent review – there is almost certainly a difference between 'full market rent' and a 'reasonable rent', the latter inferring something less than the maximum rent achievable if offered in the open market.

RICS has advised that where rental evidence is obtained from other valuers it should be provided on a consistent basis.

It is arguable that the rent agreed for a 20-year lease without review will differ from that agreed for the same lease with 10-year, seven-year or five-year reviews. Whether these rents would be higher or lower will depend upon whether the rental market is rising or falling (see Chapter 9). For preference, comparability should entail estimating rents from analyses of lettings with similar review patterns. Where this is not possible, the valuer will have to exercise professional judgment in making adjustments from the comparable evidence before using the evidence to estimate the current market rent of the subject property.

The need, in a weak letting market, for landlords to offer inducements, such as fitting out costs and rent-free periods, indicates the need for valuers to be in full possession of all the facts before reaching any conclusion on market rents being paid. For example, a rent of £50,000 pa agreed for a property for a period of five years with no rent payable for one year might be viewed by the market on a straight line basis as equivalent to a market rent of £40,000 pa:

$$( £\,50{,}000 \times 4 \text{ years}) \text{ divided by } 5 \text{ years} = £\,40{,}000.$$

The landlord's argument might be that the rental value is £50,000 a year but it has been waived for 12 months, suggesting a philanthropic landlord. However, such a genuine arrangement is possible and the valuer needs to know the full details of all rents being analysed or used as evidence. The reality is that the market will be aware of the general tone of rents at a given point in time and most valuers will know whether the £50,000 is a 'headline rent' or close to the market levels.

Other inducements can be considered in a similar way. Thus, £100,000 given for fitting out the building might be treated as equivalent over five years to a reduction of £20,000 a year, i.e. the rental value without the £100,000 would have had to fall to £30,000 a year to induce a letting.

The main issue with all such inducements is assessing the period of time to which the inducement applies. The authors' opinion is that it should apply only to the period up to the first full rent review and a discounting approach should be adopted to reflect the time value of money (see RICS Valuation Information Paper No 8, 'The Analysis of Commercial Lease Transactions' (2006)).

A current concern is that market evidence is based on quarterly, in advance rent payments, but in an increasing number of cases tenants are asking to pay rent on a monthly basis in advance. There are two issues here; the first is simply the added cost of accounting on a monthly basis, i.e. 12 times a year rather than four times a year, and the second is the question of equivalence between quarterly payments and monthly payments. Again, this can be resolved using discounting techniques. These points of principle are considered in more detail in Chapter 8 on landlord and tenant valuations.

Rental value will also be affected by a number of other lease clauses. Particular attention should be paid to those dealing with alienation – assignment and sub-letting; user; repair and service charges. Although the Landlord and Tenant Act 1988 now ensures that landlords do not unduly delay the granting of consents in conditional cases, the position in relation to an absolute prohibition on alienation remains unchanged. In the latter case there will be an adverse effect on rents. User restrictions in shopping centres and multiple tenanted commercial property can have a beneficial effect on rents, but where they are too restrictive they will have a detrimental effect on rents.

The Landlord and Tenant (Covenants) Act 1995 has amended the law on Privity of Contract. It allows landlords to define in a lease the qualifications of acceptable assignees and whilst 'reasonableness' remains a factor, some leases will be seen to be more onerous than others due to such legally permitted restrictions.

Responsibility for repairs and the nature of any service charge provisions must be considered twice: first to see if they are affecting market rents, and second to see if they leave the landlord with a liability which must be estimated and deducted as a landlord's expense before capitalisation. (For valuation purposes rent must be net of VAT, if not then the valuer will be inadvertently capitalising the future VAT liability.)

## Landlords' expenses

Where investment property is already subject to a contracted lease rent, and/or where it is customary to quote rents for a specific type of property on gross terms, it is essential to deduct landlords' outgoings (operating expenses) before capitalising the income. Investment valuations must be based on net income. This means an income net of all the expenses that the owner of an interest in property is required to meet out of the rents received from ownership of that interest in that property other than tax.

Expenses may be imposed upon the landlord by legislation, or they may be contractual, as in an existing lease. However, an inspection of the property may suggest that, though neither party is statutorily or contractually liable, there are other expenses that will have to be met by the owner of that interest in the property. The valuer must identify all such liabilities and make full allowance for them in the valuation. In order to do this, reference must be made to all existing leases in respect of the property.

The principal items of expenditure can be broadly classified under the headings of insurance, management, taxes, running expenses and repairs. Of these only the cost of complying with repairing obligations should cause any real difficulty in accurate assessment.

### Insurance

The valuer, armed with plans and his or her own detailed measurements of the building, will be able to estimate or obtain an accurate quotation for all insurances, particularly fire insurance. Fire insurance is an extremely involved subject, complicated by the variation in insurance policies offered. A valuer should therefore be acquainted with the terms and conditions of policies offered by two or three leading insurance companies and should always assess the insurance premium in accordance with those policies. If reinstatement value is required this can be referred to a building surveyor or preferably a quantity surveyor, or may be based on adjusted average figures extracted from *Spon's Architects' and Builders' Price Book* (2011). Due to the 'averaging provisions' of most policies, it is better to be over-insured than under-insured, and hence to overestimate rather than underestimate this item.

Quotations for other insurances necessary on boilers, lifts, etc can always be obtained from a broker or insurance company.

A deduction for insurance will rarely be necessary as the lease will usually contain provisions for the recovery of all insurance charges from the tenant in addition to the rent. In most cases the wording of a 'full repairing and insuring lease' leaves the responsibility for insurance with the owner, but the cost of the insurance with the tenant, the premium being recoverable as rent and hence payment is enforceable in the same way as rent.

### Management

This refers to the property owner's supervising costs equivalent to the fee that would be due to a management agent for rent collection, inspections of the premises, and instructions to builders to carry out repairs. It is frequently considered to be too small to allow for in the case of premises let to first class tenants on full repairing and insuring terms. The issue of licences to assign or sublet and to carry out alterations are often subject to additional fees payable by the tenant. The valuer must use discretion but it is suggested that if the valuer's firm would charge a fee for acting as managing agents, then a similar cost will be incurred by any owner of the property and a deduction for management should therefore be made.

The fee may be based on a percentage of rents and service charges collected or on a negotiated annual fee. The practising valuer will be aware of the appropriate market adjustments to make, based on current fees charged by managing agents. Where VAT is payable on rents it will impose an additional management cost which will be reflected in the fees negotiated by the managing agents. Where the management fee is recoverable as part of a full service charge, no deduction need be made in a valuation of the landlord's interest.

### Taxes

The Uniform Business Rate (UBR) is paid by the occupier of business premises. Inspection of the lease will indicate the party to that lease who has contracted to meet the 'rates' demand. If premises are let at an inclusive rental (i.e. inclusive of UBR) then rates should be deducted. If the property is let exclusive of the UBR then no deduction need be made from the rent. If the letting is inclusive and there is no 'excess rates clause' then the deduction must represent the average annual figure expected up until the end of the lease or next review, not a figure representing current rates. If an excess rates clause is included, then the sum to be deducted is the amount of rates due in the first year of the lease, or the lowest sum demanded during the current lease. This is because such a clause allows for the recovery from the tenant of any increase in rates over and above the amount due in the first year of the lease.

Other 'rates' may include water rates, drainage rates, and rates for environmental and other purposes, most of which are payable by the tenant in most situations.

It has always been the custom in the United Kingdom to value before deduction of tax on the grounds that income and corporation tax are related to the individual or company, and not to the property. However, if market value implies the most probable selling price, there is automatically an implication of the most probable purchaser. The valuer without this knowledge cannot be held to be assessing market value; thus it is held by some that the tax liability of the most probable purchaser should be reflected in the valuation.

For simplicity, no deduction will be made for tax in most of this book.

### Running expenses

Where the owner of an interest is responsible for the day-to-day running of a property, such as a block of flats and office building let in suites, or a modern shopping centre, a deduction from rent may be necessary to cover the cost of items such as heating, lighting, cleaning and porterage. Current practice is to include provision for a separate service charge to be levied to cover the full cost of most running expenses.

The valuer must therefore inspect the leases to check the extent to which such expenses are recoverable. Older leases tend to include partial service charges, in which case the total income from the property should be assessed and the total cost of services deducted.

All the items falling under the general head of 'running expenses' are capable of accurate assessment by reference to current accounts for the subject property; by comparison with other properties; by enquiry of electricity, gas and oil suppliers; by enquiry of staff agencies; and so on. The valuer operating within a firm with a large management department is at a distinct advantage, as that department should be able to provide fairly accurate estimates based on detailed analyses of comparable managed property.

Where a full service charge is payable by the tenant, this is usually adjustable in arrears. In other words, the annual service charge is based on last year's expenses. During periods of rapidly rising costs, the owner will have to meet the difference between service cost and service charge and reduction in the owner's cash flow and should be taken into account in a valuation.

Increasingly, these points are being met by more complex service charge clauses and schedules which provide for interim increases; for example to cover the uncertain energy element in running expenses. (The statutory requirements relating to service charges and management charges in residential property are complex. Readers involved or interested in the residential sector must refer to all current housing and landlord and tenant legislation.)

## Repairs

A detailed consideration of existing leases will indicate those items of repair that have to be met by the landlord out of rental income. The sum to be deducted from an annual income before capitalisation must be an annual averaged figure. Thus, liability to redecorate a building every five years should be estimated and averaged over the five years. A check should be made against double accounting – if the cost of repairs to boilers, lifts, etc is covered by a service charge, then no allowance should appear under repairs for such items.If indeed the cost of redecoration is recoverable by a direct proportionate charge to tenants then no deduction need be made.

Where an allowance has to be made, every effort is required to estimate the amount as accurately as possible. An excess allowance will lead to an undervaluation, inadequate allowance to an overvaluation.

Advice, if needed, should be obtained from builders and building surveyors as well as by comparison with other known repair costs for comparable managed property.

## Essential works

Any obvious immediate renewals or repairs as at the date of the valuation should be allowed for by a deduction from the estimate of capital value to reflect the cost (see Chapter 9).

## Averages and percentages

Valuers should use averages with care. An example of this is the use of average heating costs per square metre. The valuer is required to value a specific property, not an average property. The requirement is to estimate the average annual cost of heating that specific property. Some valuers, texts and correspondence courses suggest that it is a reasonable approach to base insurance premiums and repair costs on a percentage of market rent. This approach is not recommended unless the valuer can prove that the figure adopted is correct. A percentage allowance may be widely inaccurate; a high street shop could be let at £100,000 a year and a factory could be let at £100,000 a year, but the latter will almost certainly cost more to keep in repair. Two properties may let at identical rents, but one may be constructed of maintenance-free materials, the other more cheaply constructed with short-life material. A thousand square metres of space in Oxford Street, London, and in Exeter, Devon, may cost the same to maintain but could have very different market rents. A large old building may let at the same rent as a small modern building, and clearly the repairs will differ.

Similar points may be made in respect of the use of percentages to estimate insurance premiums. Students may be permitted to use percentages in class examples if the data needed for accurate assessment is not provided. Say 0.5% to 1% of MR for insurance, 10%

of MR for external and internal repairs, possibly split 6% and 4% between external and internal, 1% to 5% of MR for management and 10% to 15% as the fee for managing a full service charge for, say, a shopping centre. Students must reflect on the reality of the figures they calculate when using percentages.

### Voids

Where there is an over-supply of space within a given area the probability of voids occurring when leases terminate is increased. If a single investment building is let in suites to a number of tenants, voids may occur sufficiently regularly for a valuer to conclude that the average occupancy is only, say, 80%. In such cases, having estimated the market rent, the figure should be reduced to the level of the most probable annual amount. Voids are less likely to affect operating expenses so a pro rata allowance on operating costs should not be made. Indeed in some exceptional cases it may have to be increased to allow for empty rates, non-recoverable service costs and additional security.

In the case of a building let to a single tenant, a void is only likely to occur at the end of the current lease or if there is a break clause. If this is a reasonable expectation, then the income pattern must be assumed to be broken at the renewal date for the length of time considered necessary to allow for finding a new tenant. Additional allowances may have to be made for negative cash flows if empty rates and other running expenses have to be met by the owner during this period.

The time taken to renew leases to current tenants and to negotiate rent reviews can represent a significant loss to a landlord. New leases will contain 'interest' clauses to cover interest lost on rent arrears, and delayed settlement of reviews and renewals.

In other cases the valuer may wish to adjust rental income down to reflect the fact that agreement is unlikely to coincide with the rent review or lease renewal date.

## Purchase expenses

### SDLT, solicitors' fees, etc

It has long been valuation practice when giving investment advice to include in the total purchase price an allowance to cover stamp duty, solicitors' fees, and any other expenses to the purchaser occasioned by the transaction including VAT. This in total is currently assumed to come to 5.75% of the purchase price, although valuers tend to introduce a degree of precision into their art by using 5.7625%, where SDLT is 4% on sales of commercial property over £500,000 and legal fees (contract and conveyance) plus charges, plus surveyor, valuer and environmental survey fees might amount to 1.5% and VAT which is payable on fees but not on charges or SDLT. These may come to 5.7625%, but for many investors with internal teams of lawyers and surveyors it might be as little as 4% plus 0.5%. The increase of VAT to 20% will move this standard to a convenient 5.80%.

It should be remembered that analysis of a transaction might also take these expenses into account to reveal the buyer's yield on total outlay rather than the yield realised on the purchase price alone.

It is current practice in investment valuation work to express value net of these sums. However, MV is normally expressed without deduction for selling expenses.

## Income capitalisation or DCF

The first step in the income approach is to identify the actual net rent being paid (net operating income or NOI in North American appraisal practice) and the current market rent in net terms.

The second step is to determine the most appropriate basis of discounting the income; that is, whether to use:

- Non-growth DCF.
- Growth explicit DCF.
- Conventional income capitalisation.
- Rational or real value methods.

In an active property market, the valuer will be able to determine an appropriate capitalisation rate (all risks yield) from the analysis of sales of comparable properties.

For a property to be comparable it will need to be:

- similar in use
- similar in location
- similar in age, condition, design, etc. (i.e. to have similar utility value)
- similar in quality of tenant
- let on similar lease terms.

Where a property sale meets most but not all of these criteria, the experienced valuer with good market knowledge will be able to adjust the capitalisation evidence obtained from the market for any variations.

Experience or knowledge here means knowing:

- the current players in the market
- who is selling which type of property and why
- who is buying which type of property and why
- the current state of the market and the underlying reasons for it rising or falling
- the national, regional and local economic situation.

Analysis of current sale prices must be undertaken on a consistent basis and the valuer needs full details of the transaction in order to assess the capitalisation rate net of purchase costs.

**Example 4.5**

An office building is currently let at its market rent of £100,000 a year net of all outgoings, with rent reviews every five years. It has just been sold for £1,250,000.

Assess the capitalisation rate.

$$\frac{100,000}{1,250,000} \times 100 = 8\%$$

In practice, the transaction may have been completed at a price to produce a return of 8% on total acquisition cost to the purchaser and thus £1,250,000 would represent price plus fees hence:

let purchase price = £x
let purchase costs = 5.75%

then 1.0575x = £1,250,000 and £1,250,000 ÷ 1.0575 = £1,182,033.
and capitalisation rate = 8.46%.

A variation in capitalisation rate of 0.25% can be significant at the lower rates.

Thus:

| Income | £100,000 |
|---|---|
| PV £1 pa in perp at 4.50% | 22.22 |
| | £2,222,000 |

| Income | £100,000 |
|---|---|
| PV £1 pa in perp at 4.00% | 25.00 |
| | £2,500,000 |

Here, the 0.50% variation represents £278,000 or a variation in value of 11.12%

The market evidence and market knowledge will allow the valuer to express an opinion on capitalisation rates with some confidence and to be able to make 'intuitive' or professional adjustments for variation between the subject property and the market evidence. If the market evidence is 8% and the valuer is to use that to value a comparable but marginally superior investment property, the capitalisation rate will be adjusted down to less than 8%. If the property being valued is marginally inferior, the rate will be adjusted to more than 8%. There is no rule, but on a point of principle one might argue that any adjustment between comparable and subject property in excess of 0.25% might suggest the comparable is in fact not comparable.

In some cases, the property will display qualities which add to and detract from the quality of the investment. Thus, a property may be in a marginally better location but may be marginally less efficient having only 75% net to gross floor area compared to the normal 80% for the market and so the adjustments cancel out.

The subtleties of market intuition cannot be taught, they have to be acquired from experience.

Where the property sold is producing a rent (income) below MR, and there is an expectation of an early reversion, the capitalisation rate can be found by trial and error or iteratively using a financial calculator or, if Excel is used, 'Goal Seek'. This is an IRR calculation and the rate percent found is called an equivalent yield.

### Question 1:

A freehold shop let at its open market rental value of £75,000 a year with rent reviews every 5 years has just been sold for £1,250,000. Assess the market capitalisation rate.

### Question 2:

A freehold office let at £30,000 a year with 2 years to the next rent review has just been sold for £553,000. The current market rent is £40,000. Assess the equivalent yield by trial and error.

## DCF

All income capitalisations are a simplified form of DCF. Thus, the capitalisation of £100,000 a year in perpetuity at 8% can be shown to be the same as discounting a perpetual (forever) cash flow of £100,000, receivable at the end of each year in perpetuity.

Nevertheless, there are growing arguments for using a cash flow approach for all income valuations – in particular a DCF approach. This:

- provides clearer meaning to investors, bankers and professional advisors
- can incorporate regular and irregular costs such as fees for rent reviews, lease renewals, voids, etc
- can allow for expected refurbishment costs
- can incorporate adjustments for obsolescence.

DCF is the normal basis in North America, where property is frequently let on gross rents with provision for landlords to recoup increases in certain service costs but not all.

Additionally, DCF methods may be a preferred approach when dealing with new developments which are not expected to achieve full potential for several years following practical completion.

The current use of the term 'DCF' by UK valuers generally refers to the use of a modified DCF, (rational method) or real value approach to valuation. Some valuers believe that these methods should be used for all valuations, whilst others see their use being reserved for those properties which display an income pattern which is out of line with the normal expectations in the market place.

The use of DCF and conventional capitalisation approaches are considered in Chapters 6 and 7 for the valuation of freehold and leasehold interests in property.

# Summary

The key steps in the income approach to property valuation are as follows:

- Agree purpose of valuation, basis of value and other terms and conditions with the client.
- Assess current net operating income.
- Assess income potential based on comparable evidence of current lettings.
- Assess appropriate capitalisation rate.
- Consider use of cash flow layout in order to incorporate specific allowance for fees, depreciation, etc.
- Complete the valuation and prepare a full report.
- In any valuation falling within the requirements of the RICS Valuation Standards the valuer, if a member of the RICS or IRRV, must comply with those standards.

# Chapter 5

## The Income Approach: Freeholds

## Introduction

On receipt of instructions to value property a valuer, if a member of the Royal Institution of Chartered Surveyors (RICS), is required to confirm the terms of engagement (RICS Valuation Standard (VS 2); the majority of valuations are now required to be compliant with the RICS *Valuation Standards* (The Red Book) (2011) which are themselves compliant with the International Valuation Standards (IVS). There are a few exceptions set out in VS 1.1, but the underlying assumption in this text is that the requirements of The Red Book must be followed.

In agreeing the terms of engagement, the valuer will identify the client, the purpose of the valuation, the subject of the valuation and the interests to be valued. The main legal interests that are bought and sold in property are freehold and leasehold interests. As investments, the former have an assumed perpetual life because the freehold title (fee simple absolute) is the largest legal estate or right of ownership of land in the United Kingdom, a title which enjoys rights in perpetuity; leasehold interests have a limited life span fixed by the lease term and limited rights of enjoyment as specified in the lease covenants.

Having agreed the terms of engagement, the valuer will then complete all necessary investigations in accordance with VS 5 and to the extent as agreed in the terms of engagement. This is the fact finding stage of the process. In terms of investment property, that is property tenanted and held as an investment, all leases will normally be read in order to establish what current benefits, by way of rent, ownership will produce for an investor and to assess any liabilities that the owner may be, or may become, liable for in the future. The valuer must then assess the market rent (see chapter 4). Having determined both the current rent passing and the market rent, net of any landlord's annual expenditures on repairs and insurance, the valuer is able to judge whether the property is let at market rent, let below market rent or let above market rent. The timing of these potential receipts and payments and other factors will also be established from the lease, i.e. when was it granted? For how long? When will it terminate by effluxion of time? Are there any rent reviews? How frequent are they and when is the next one due? What is the basis of the rent review? Is the lease on a turnover rental basis or index linked?

The other key element in an income valuation is the yield. Valuers and their research teams monitor market activity, both in terms of volume of trading and in terms of the yields being achieved on the sale of investment property. This monitoring will indicate whether the market is stable, rising or falling. The most frequently reported market reports relate to the price of residential property, where the popular press is regularly reporting that year on year, or month on month, prices have risen by x% or fallen by y%. The investment market is no different in this respect as it too shows periods of price rises, and periods of price falls. This information is important because the valuer is seeking to estimate market value today on the basis of known sale prices that have occurred. There may be a time lag between the most recent sales and the date of valuation. The valuer must reflect on market movement and the factors causing those movements and make a judgment as to whether the market is going to continue to rise or fall. Investment consultants will take this analysis much further, as their concern is not with a snapshot, but with the future uncertainties that affect any decision to buy (see Chapter 12).

The yields themselves are determined from analysis of prices obtained; this is, in effect, an assessment of the internal rate of return (see Chapter 3) as at the date of sale but excluding any explicit growth in rents. This is straightforward when the property is let at the market rent and can be found from:

$$\frac{Income}{Saleprice} \times 100 = yield$$

So, a property let at £113,000 and sold at a contract price of £1,500,000 will be analysed as:

$$\frac{113,000}{1,500,000} \times 100 = 7.53\%$$

The valuer may analyse prices before or after allowing for purchase costs but must be consistent between analysis and valuation. To infer that buyers and sellers of property are aware, in an imperfect market, of perfect data such as 7.53% is probably unrealistic; this one piece of data will be weighed with other data and a view taken that property of a similar type, at this time, in the market, might well sell at yields of around 7.50%. The valuer is continually adjusting from the known to the unknown in the light of a plethora of market and property specific data which may suggest that the yield to be used for the similar property to be valued should be adjusted up or down. Adjustment for market movements, as illustrated in Figure 5.1, is one of the most demanding but it has to be supported with reasoned argument.

A market valuation is at the date of valuation not at the date of historic market evidence, but the market value must be based on market evidence; this will also be the case in estimating market rent - are rents on the rise and continuing to rise, or falling and continuing

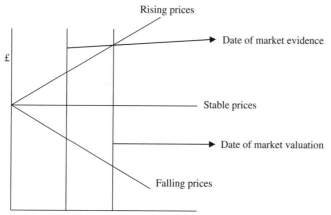

**Figure 5.1** Adjusting for market conditions

to fall, and what is the evidence to support the valuer's conclusion. There may be enough market sentiment to support the valuer's conclusion but it will be the valuer's judgement as to the degree of rise or fall that has occurred. At this stage it should be noted that if a fall in prices has occurred and yields are hardening by 0.25%, it will mean a price move for a similar property from £1,500,000 to £1,458,000 say, in the scale of values, £1,450,000 and if prices are rising and yields are softening by 0.25% then comparable values will move to £1,558,000 say £1,560,000.

A yield analysed in this way may be called a capitalisation rate or initial yield. After consideration of the various income approaches to valuation which are used by valuers for freehold valuation, the analysis of sale prices for properties which are let at less than market value will be considered.

The market is generally analysing the sale prices of property where the rent is paid quarterly in advance, but custom and practice is to analyse on the assumption that income (rent) is received annually in arrears. This means that the yield is not a true yield but a nominal yield. Again, it is important for analysis and valuation to be undertaken on a consistent basis. It would be wrong to use annual in arrears yields data and then to value on the basis of quarterly in advance.

Most software packages used by valuers for investment valuation work will provide the valuer with the yield and the effective yield referred to by valuers as the True Equivalent Yield (TEY). A yield of 7.5% on a quarterly income basis is a TEY of 7.8652%.

If £1,333.3334 is paid for an income of £100 (7.5% nominal) but in fact £25 is paid quarterly in advance then only £1,308.3334 has been paid for £25 per quarter in arrears and so:

$$\frac{25}{1,308.3334} \times 100 = 1.910828\% \text{ per quarter which is } 1.01910828^4$$
$$= 1.0786519$$

*deduct the initial 1 to get 0.0786519 and multiply by*
*100 to convert to a percentage of 7.86519 say 7.8652%*

# The income approach

The perpetual nature of freehold interests in property means that valuers have developed certain short cuts, as clearly it is not sensible to try and look too far into the future.

Two distinct approaches have developed for assessing the market value of an income-producing property. The first is to assume a level, continuous income flow and to use an overall or all-risks capitalisation rate derived from the analysis of sales of comparable properties let on similar terms and conditions (i.e. five-year or seven-year rent review patterns) to calculate present value, that is market value.

The second has been named the discounted cash flow or DCF approach.

The first or normal approach is favoured by many on the grounds that it is more correctly a market valuation. As such, it relies on an active property market and an ability to analyse and obtain details of comparable capitalisation rates. The definition of market value implies that there must be a market for the asset being valued. If not, then the necessary ability to compare prices achieved to support an opinion of value is lost; there is no market so there can be no market value. However, those in favour of the DCF approach argue that, when there is limited property market activity, a valuation can still be undertaken using a DCF approach. Market activity in this respect refers not only to the total volume of sales at a point in time, but a sufficiency of direct comparables. When there is little evidence for direct comparison, or in the case of the income approach, indirect comparison, a view as to what the value should be may be derived indirectly from other market activity and developed through the DCF approach.

For example, it is assumed that if initial yields on market rented properties are, say, 7% then the capitalisation rates to be used for valuing property with an early reversion to a market rent can be closely related to 7%. Further, if there is a long period of time to the reversion then the 7% can simply be adjusted upwards for the change in risk by the valuer. The argument against this is that if there is no evidence of capitalisation rates from sales of investments with early reversions or long reversions it may be because there is no market for such investments. However, in order to support an opinion of market value, required by a client, the DCF approach, which is built up from the known market evidence, is more supportable and likely to be more accurate than any intuitive adjustment made by a valuer.

The strongest criticisms of the normal approach are that it fails to specify explicitly the income flows and patterns assumed by the valuer and that growth implied all risk yields are used to capitalise fixed flows of income. The DCF approach requires the valuer to specify precisely what rental income and expenses are expected, when and for how long. The valuer is therefore forced to concentrate on the national and local economic issues likely to affect the value of the specific property as an investment. There may after all be properties in a depressed economy for which rental increases in the foreseeable future are very unlikely.

The DCF approach accepts the idea of the opportunity cost of investment funds. Opportunity cost implies that a rate of return must be paid to an investor sufficient to meet the competition of alternative investment outlets for the investor's funds. This is the basis of the risk free rate of discount, a risk free rate being assumed to compensate for time preference only. Any investment with a poorer liquidity factor or higher risk to income or capital value will have to earn a rate over time in excess of the risk free rate. Analysts tend to adopt, as their measure, the current rate on 'gilt edged' stock (normally issued by governments and therefore their marketability is dependent upon international perceptions on the strength of a government and its policies). These are generally held to be fairly liquid investments and are safe in money terms if held to redemption. Additionally, if they are sold at a loss, or at a gain, it can be reasonably assumed that there will have been a similar movement in the values of most other forms of investment.

Property is expected to achieve a return, over time, of 2 to 4% above the going rate on gilt edged stock, or higher in the case of poor quality properties or locations. This is because it is less liquid, indivisible, fixed in location, costly to manage, dependent on credit and therefore generally harder and more costly to sell. This property risk premium is not a 'given', as under certain market conditions demand can be sufficiently strong to force prices up and yields down to the level of stock redemption yields or below. The valuer has to appreciate the full investment market not only to be able to value and to reflect the changing market conditions, but also to be able to advise investors on the appropriateness of investing in property in general and in specific properties in particular. The DCF method requires the valuer to 'forecast'; that is to specify the expected rents at due dates in the future and to discount those rents at a rate sufficiently higher than the risk free rate to account for the additional risks involved in the specific property investment. If there is no risk premium then the use of the DCF approach will make this fact explicit. 'Forecast' here does not mean prediction nor does it necessarily imply a projection based on extrapolating or extending the past into the future. It is an estimate of the most probable rent due in 'n' years' time, based on sound analysis of the past and present market conditions. One approach is to assess an implied growth rate for rent from the relationship between an all risk capitalisation rate and the hurdle or target rate that investors require to compensate for the extra risk of property compared to government stock.

The 2010 RICS Guidance note, *Discounted Cash Flow for Commercial Property Investments*, is now part of the 2011, seventh edition of The Red Book (GN7). It contains important advice for valuers using an explicit DCF valuation method. It recognises that DCF can be used to estimate market value and that in the absence of transactions it provides a rational framework for estimating market value. As described here, and elsewhere in this book, DCF is implying that a market price is one which rational investors will be willing to pay, and can be deduced by comparison with the returns from other investments having regard to their comparative risks. The RICS advice covers issues such as: estimating the cash flow and the need to avoid double counting; estimating the exit value where one may be seeking 'to hone the exit yield into a long-run equilibrium yield that provides a measure of longer stability by absorbing shorter-term volatility and uncertainty'; selecting the discount rate, 'the discount rate (the target rate of return) is usually derived by reference to the return on an alternative form of perceived low-risk or risk-less asset (frequently the benchmark is the gross redemption yield on government gilts or cash), plus appropriate additions for risk', with an interesting section considering the need to look back and to look forward, noting that here the risks relate to market factors and property specific factors and one of the key issues is the difficulty of future structural change – thought must be given to what might be a subjective consideration of what might happen.

The RICS guidance is now a 'must read' and emphasises the importance of distinguishing its use for assessment of investment value or worth and the assessment of market value.

Typical capitalisation and DCF approaches are set out below in relation to standard market valuation problems. Direct capitalisation is seen to be a preferred approach, but practice increasingly sets this out in a cash flow format and may supplement this with a 'what if' growth explicit DCF. The preferred approach of those involved in property development and in property investment is DCF and it is now growing in use in the area of property valuation.

## Capitalisation approaches

The technique of converting income into a capital sum is extremely simple. In the case of freehold interests in property, the income will have the characteristics of an annuity which may be fixed, stepped, falling or variable.

### A fixed or level annuity

If the income is fixed for a period much in excess of 60 years or in perpetuity, or if a property is let at its market rent and there is market evidence of capitalisation rates, then it can be treated as a level annuity in perpetuity, see Figure 5.2.

$$\text{Net income} \div i = \text{Capital value}$$

$$\text{Or net income} \times \frac{1}{i} = CV$$

---

### Example 5.1

A shopping centre was developed on a ground lease. The ground lease is for 125 years at a ground rent of £100,000 a year. The lease has no break clauses or rent review clauses. The lease has 90 years to run.

- Fixed income of £100,000
- Unexpired term 90 years as at valuation date
- Capitalisation rate 10% based on market comparables

£100,000 ÷ 0.10 = £1,000,000
A conventional valuation would appear as:

| Net income | £100,000 |
|---|---|
| PV* £1 pa in perpetuity at 10% | 10 |
| (YP** in perpetuity at 10% 1/i) | |
| Estimated market value | £1,000,000 |

*Present value.
**Years' purchase.

A perpetuity approach is used because the PV £1 pa factor for 90 years is equivalent to that in perpetuity (9.9981 cf. 10.00).

There may be market comparables to support the capitalisation rate of 10%. However, in the absence of evidence, the rate can be compared to that achieved on irredeemable government stocks with an adjustment for any extra risks or security attaching to the property investment. Investments of this nature are very secure in money terms but not in real terms. The money security is provided by the fact that the ground rent is secured by the rents paid by the occupying tenants to the developer and by the fact that the ground lease would be forfeited in English Law on non-payment of the ground rent of £100,000.

A similar approach can be adopted for properties currently let at their market rent.

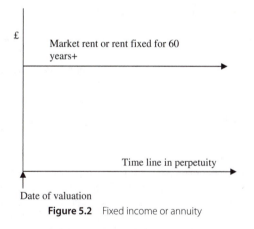

**Figure 5.2** Fixed income or annuity

### Example 5.2

Value a freehold shop in a prime trading position let at its market rent of £120,000 a year on full repairing and insuring terms (FRI), meaning that the rent is all net investment income to the owner. The lease is for 15 years with upward only rent reviews every five years. No break clauses. Similar properties are selling on the basis of an all risks yield (ARY) or capitalisation rate of 5.5%.

| Rent | £120,000 |
|---|---|
| PV £1 pa (YP) perpetuity at 5.5% | 18.1818 |
| Market value | £2,181,818 |

Notes:

1. Where the lease is for 15 years but it can be assumed that it would continue to let readily and that market rents will always be receivable then it can be treated as a perpetual income. Such an assumption could not be made if the trading position was under threat from planned development such as an out-of-town centre, or where the tenant's financial position is not secure.

2. Although the rent is reviewable every five years, there is no need to provide for this explicitly in a capitalisation approach if at each reversion the rent would also be equivalent to today's market rent of £120,000. The key distinction between Example 5.1 and 5.2 is that the former is actually fixed for 90 years, whereas the latter is inflation proofed by the rent reviews.

3. All property incurs a management cost but customarily when property is let on full repairing and insuring (FRI) terms no deduction is made for management.

4. For some purposes the costs of acquisition (survey fees, legal fees and Stamp Duty Land Tax (SDLT)) will be deducted to arrive at a net value. Where the net value plus fees adds up to £2,181,181 then £120,000 will represent a return on total outlay of 5.5% (see Chapter 12 for more information on property investment analysis and advice). The return on net value would be greater than 5%.

### Example 5.3

If the rent in Example 5.2 is payable on an annual, in advance basis and the market is still looking for a 5.5% investment, the valuation becomes:

| | |
|---|---:|
| Rent | £120,000 |
| PV £1 pa in perp at 5.5% in advance | 19.1818 |
| Market value | £2,301,818 |

Notes:

1. £2,301,818 is being paid for the right to £120,000 immediately and thereafter £120,000 every year at the end of the year. A lease providing for payments annually in advance is very rare. This value would only occur if the valuation was undertaken on the rent due date. In such a case it is not the terms of the lease but the date of the valuation and the rent payment date that guides the valuer. In property sales, the contract will usually provide for rental apportionment as between seller and buyer. The definition of market value implies contractual completion at the date of valuation. The in advance rent would be paid to the seller by the tenant and transferred to the buyer as part of the contract. Hence, the extra £120,000 is paid to receive the immediate £120,000 (£2,301,818-£2,181,818 = £120,000).

2. Similar points and changes can be made to reflect quarterly rental arrangements where the PV £1 pa in perpetuity quarterly in advance would be used, but switching to quarterly in advance requires the yield to be based on quarterly in advance analysis not annual in arrears analysis (see Chapter 2).

Valuers are being encouraged to represent their investment valuations as cash flows as this is a more familiar format for accountants and investment analysts.

| Period | Net Income | PV £1 at 5.5% | NPV |
|--------|-----------|---------------|-----|
| 1 | 120,000 | 0.94787 | 113,744 |
| 2 | 120,000 | 0.89845 | 107,814 |
| 3 | 120,000 | 0.85161 | 102,193 |
| 4 | 120,000 | 0.80722 | 96,866 |
| 5 | 2,301,818* | 0.76513 | 1,761,190 |
| | | | £2,181,807 |

*(£2,301,818 represents the resale value at the end of year 5 being the right to continue to receive £120,000 pa thereafter in perpetuity plus the rent due at the end of year 5 of £120,000; the small variation is due to rounding in the table of figures, Excel would be used for accuracy.)

This structure, set out in a spreadsheet, would allow the valuer to revalue using a range of rates where there may be some degree of uncertainty as to the market rates. The structure can also be lengthened and developed to reflect growth.

## A 'stepped' annuity

If the income is fixed by a lease contract for 'x' years and is then due to rise, either by reversion to a market rent or to rise to a higher level for 'y' years, then reverting to a market rent in perpetuity, the valuation may be treated as an immediate annuity plus a deferred annuity in one of two ways. The first is referred to as a term and reversion, the second as the layer method.

Figures 5.3 and 5.4 (below) illustrate the distinction between these two approaches; the first splits the income (rent) between the current rent being paid and the future rent at rent review or lease renewal, the second – the layer approach – splits the income (rent) between that which can be assumed to continue in perpetuity from the date of the valuation and the extra rent that will be received following the rent review or lease renewal. If the same capitalisation rate or equivalent yield is used throughout both approaches then the values will be the same. Choice is a matter of preference; term

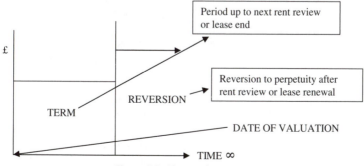

**Figure 5.3**   Term and reversion

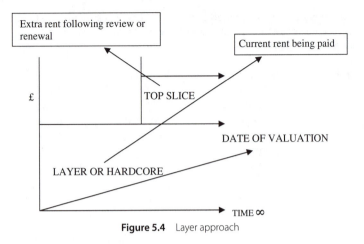

**Figure 5.4**   Layer approach

and reversion may be simpler if outgoings have to be deducted or
an allowance for a void has to be made (see Example 5.5) and the
layer is simpler for over-rented property (see Example 5.8).

### Example 5.4

If the property in Example 5.2 is let at £100,000 pa on FRI terms with two
years to run to a review to the market rent of £120,000, the valuation
could appear as:
Term and reversion

|  |  |  |
|---|---|---|
|  | £100,000 |  |
| PV £1 pa (YP) 2 years at 5.5% | 1.8463 | £184,630 |
| Reversion | £120,000 |  |
| PV £1 pa (YP) perp. at 5.5% 18.1818 |  |  |
| x PV £1 in 2 years at 5.5% 0.89845 | 16.3355 | £1,960,260 |
|  |  | £2,144,890 |

Layer

|  |  |  |
|---|---|---|
| Layer (or bottom slice) | £100,000 |  |
| PV £1 pa (YP) perp. at 5.5% | 18.1818 | £1,818,180 |
| Top slice | £20,000 |  |
| PV £1 pa (YP) perp. at 5.5% 18.1818 |  |  |
| x PV £1 in 2 years at 5.5% 0.89845 | 16.3355 | £326,710 |
|  |  | £2,144,890 |

Notes:
1. Because all the capitalisation and deferment or discounting rates
   have been at 5.5%, both conventional methods produce the same
   value estimate.
2. The £100,000 is the rent currently being paid; two further in arrears
   payments are due and have to be discounted to present value at

5.5%. However, in the layer the £100,000 is assumed to continue in perpetuity as it will be the first part of the reviewed rent of £120,000.

3. The reversion in the term is today's market rent even though the actual reversion is two years away; this is the conventional approach, the yield reflecting the markets expectation of any market movements in rent between today and two years time. In the layer the reversion is to the additional rent expected of £20,000.

4. The reversion and top slice are assumed to be in perpetuity, capitalised at the equivalent yield and then deferred for the two year period of waiting.

Here again a cash flow would represent this as:

| Year | Net Income | PV £1 at 5.5% | NPV |
|------|-----------|---------------|-----|
| 1 | 100,000 | 0.94787 | 94,787 |
| 2 | 100,000 | 0.89845 | 89,845 |
| 3 | 2,181,818 | 0.89845 | 1,960,255 |
| | | | £2,144,887 |

And once more rounding to five decimal places accounts for the small variation.

Some valuers still insist on using different rates to reflect some personal view on the security of the current rent and the additional risk of the future rent. This can be dangerous and difficult to support. This is shown below for both approaches.

Term and Reversion

| Term | £100,000 | |
|------|----------|---|
| PV £1 pa 2 years at 4.5% | 1.8727 | £187,270 |
| Reversion | £120,000 | |
| PV £1 pa in perp def'd 2 years at 5.5% | 16.3355 | £1,960,260 |
| | | £2,147,530 |

| Layer | | |
|-------|----------|---|
| Bottom slice | £100,000 | |
| PV £1 pa in perp. at 5.5% | 18.1818 | £1,818,180 |
| Top slice | £20,000 | |
| PV £1 pa in perp def'd 2 years at 6.5% | 13.5640 | £271,280 |
| | | £2,089,460 |

The two arguments are different. In the term and reversion it is being suggested that a tenant paying below market rent is less likely to default than one paying market rent; the rent is more secure so a lower capitalisation rate is used to reflect the reduced risk for two years. The reversion remains at the market capitalisation rate for this quality of property investment. In the layer great care is needed to avoid falling into the trap of reducing the rate for the bottom slice, or continuing income, on the same arguments of security; to do so would suggest it is more secure forever, whereas it is clear that in two years it will be at the normal market risk. The adjustment that might be accepted is about gaining the extra £20,000 in two years time, it is not contracted and so there is an argument to place this expectation at a higher risk. The impact is greater. The argument against any adjustment is that using an equivalent yield the valuer is reflecting all the risks already in the choice of market yield. To start adding elements of risk to parts of the income can confuse the issue as shown below.

The reason for valuers wishing to change rates within a valuation rests with the historic evolution of the methods and the changing economy as reflected in the investment market (see Baum and Crosby (2008)). In the immediate, post credit crunch market readjustment, the argument for separating contracted income from hoped-for income became very real. This added risk factor pushed yields higher as a result. So further care is needed to ensure that the risk is not being counted twice; once by increasing the equivalent yield and then by raising the yield on the top slice; but due diligence in assessing the tenant risk is paramount in any valuation.

The reason for yield adjustment within a valuation probably owes its origin to valuation practice in a very different economic climate. Valuers in the nineteenth century were primarily dealing with rack rented properties, largely agricultural, at a time when inflation was relatively unknown. A realistic and direct comparison could be made between alternative investments such as the undated securities (gilts) issued by government, the interest on deposits and farmland. The latter, being more risky, was valued at a percentage point or so above the secure government stock. Other property, such as residential and the emerging retail properties, were more risky but were let on long leases so they could be capitalised in perpetuity, again at an appropriate higher rate.

In the 1930s, the depression made any contracted rent a better or more secure investment than an empty property, and there was real fear that at the end of a lease the tenant would seek to redress the position by requesting a lower rent or by vacating. Logic suggested the use of a lower capitalisation rate for secure contracted rents than unsecured reversions.

This approach continued unchallenged until the 1960s. In the meantime the market had changed in many ways and, with the rise of inflation, the investment market had identified a crucial difference between fixed-income investments such as gilts, and those where the owner could participate in rental and capital growth such as equities and rack rented properties.

This change was noted in the market and the reverse yield gap emerged. Nevertheless, valuers continued to use a capitalisation approach, varying yields to reflect the so-called security of income of contracted rents, the argument being that a tenant would be more likely to continue to pay a rent which was less than the estimated market rent because it represented a leasehold capital value (see Chapter 6). The market did not appear to recognise that the technique of the 1930s was one of using rates to reflect money risk, whereas those after the appearance of the reverse yield were capitalisation rates which reflected money risks and expectations of growth.

There are currently many arguments for not varying capitalisation rates in simple conventional valuations, but two will suffice. By the 1960s, the property market for income-producing properties was being dominated by the major institutions and property investment surveyors were increasingly required to specify in their reports:

- the initial or year one rate of return
- the investment's internal rate of return (IRR).

The former acted as a cut-off rate as actuarial advice at that time required all investments, depending upon the fund, to produce a minimum return – frequently 3 to 4%. It still acts as a cut-off rate as some investors will not buy unless the initial yield is above an actuarially specified minimum. The second caused confusion for valuers used to using perhaps four or five rates in a multiple reversionary property. The confusion arose because few valuers understood the concept of IRR, few knew how to calculate it, and in the 1960s there was nothing more sophisticated than a slide rule and logarithm tables to help with the calculations. The profession took a little longer to recognise that if the capitalisation rate used in the valuation was held constant it would be the investment's expected IRR based on current market rents.

It can also be shown that in most term and reversion exercises where variable rates are used the IRR – now popularly called the equivalent yield – is almost the same as the ARY. As a result many valuers, whether using the term and reversion or layer method, now use an equivalent or same yield approach. It is, however, important to distinguish between the terms 'equivalent' yield and 'equated yield'.

In Donaldson's Investment Tables it is stated that:

> The equated yield of an investment may be defined as the discount rate which needs to be applied to the projected income so that the summation of all the income flows discounted at this rate, equals the capital outlay;... whereas an equated yield takes into account an assumed growth rate in the future annual income, an equivalent yield merely expresses the weighted average yield to be expected from the investment in terms of current income and rental value, without allowing for any growth in value over the period to reversions.

Finally, it can be shown that varying the rates between tranches of income can pose its own hidden problem. Bowcock (1983) uses a simple example to illustrate the danger. He uses two identical properties, each let at £100 a year but one has a rent review in five years and the other a review in 10 years, both have a reversion to a market rent of £105 a year.

Traditionally, the solution might appear as:

| | | |
|---|---|---|
| Term | £100 | |
| PV £1 pa 5 years at 9% | 3.8897 | £388.97 |
| Reversion | £105 | |
| PV £1 pa perp def'd | | |

| | | |
|---|---|---|
| 5 years at 10% | <u>6.2092</u> | <u>£651.96</u> |
| | | £1040.93 |
| | | |
| Term | £100 | |
| PV £1 pa 10 years at 9% | <u>6.4177</u> | £641.77 |
| Reversion | £105 | |
| PV £1 pa perp def'd | | |
| 10 years at 10% | <u>8.8554</u> | <u>£404.82</u> |
| | | £1046.59 |

Clearly, there is something wrong with a method which places a higher value on the latter investment, which includes a 10-year deferred income, than the former with a five-year deferred income. In practice this error is not normally noted by valuers, who subjectively adjust their rates to reflect (a) their view of the extent of the security of income, namely the difference between contracted rent and market rent, and (b) the risk associated with the period of waiting up to the rent review. They do not compare to see the effect of the changes on their opinion of value.

Some critics have also drawn attention to the problems of selecting the correct deferment rate in a variable yield valuation. Customarily, the reversion has been deferred at the reversionary capitalisation rate. If this is done then it can be shown that the sinking fund provision within the YP single rate formula (at 9% in the above example) is not matched by the discounting factor (10% in the example); as a result, the input rates of return of 9% and 10% will not be achieved. This can be corrected by deferring the reversion at the term rate and is avoided using an equivalent yield.

The discerning valuer should conclude that it is safer and more logical to adhere to the same yield or equivalent yield approach. It is also the easiest yield to extract from sales evidence as the calculation is of the IRR. On a same yield basis there is no distinction to be made between the two capitalisation methods. However, the term and reversion seems the most acceptable theoretical method and preferable for valuations involving lease renewals when a void allowance or refurbishment cost may need to be built in. It is also easier to handle if outgoings have to be deducted. The layer method is possibly simpler and more useful when handling certain investments where clients wish to know what price has been or is to be paid for the top slice. An example would be the valuation of over-rented property where the top slice might cease at the lease renewal when the rent reverts to the lower market rent.

Valuers are still faced with the problem of finding comparables. The more unusual the patterns of income, the more difficult it is for the valuer to judge the correct capitalisation rate. For example, what would be the right equivalent yield to use for a property let at £100,000 a year with a reversion to £120,000 a year in 15 years?

The problem in these cases is that the period to the reversion is beyond the normal rent review or lease length, without any opportunity for the rent to be adjusted for inflation or growth. The investment is more risky in money terms – it is not inflation proof. In these cases an equivalent yield approach is likely to be criticised if the yield used is taken direct from the typical market and not from a comparison with similar long reversionary properties. The market solution is to intuitively adjust the capitalisation rate upwards. The question the valuer has to decide is how big should the upward adjustment be, and does it increase with the length of the reversion or with the scale of the reversion or with both. For example, should a different view be taken of a reversion in 15 years where the increase in rent will be, in today's terms, 100% to one where the increase will be 500%?

The position during periods when there is a positive reverse yield gap, as there was in the 1980s/1990s was one where the DCF approach or real value approach could be used to advantage (see page 108). This is still the case where investments with growth potential are selling on a low yield basis compared to the fixed income returns on government stock, plus an adjustment for the added risks associated with property. In such cases the risk position has not changed, as all property is still riskier than holding cash or gilt edge stock. Thus, if gilts are at say 12%, as they once were, a property investor should arguably be looking for 13 to 14%. Purchasing at 5% is forgoing at least £8 per £100 invested each year. Few people, if offered employment at £30,000 a year, would reject it in favour of an alternative at £20,000 a year unless they were certain that over a specified contract term the loss of £10,000 a year would be compensated by regular and substantial increases in salary. The same is true of property. So, capitalising at 5% when redemption yields on gilts or investors target rates are substantially higher must imply an expectation of rental growth in perpetuity, and the first criticism of the equivalent or same yield approach for reversionary investments is that it fails to indicate rental growth explicitly, the reversion is always to a rent expressed in market terms as at the date of valuation.

It should also be noted that the term and the reversion are capitalised at the same ARY, implying the same rental growth, when clearly a lease rent is fixed by contract for the term. The substantive argument in defence of the equivalent yield approach is that of simplicity. The modified or shortcut DCF and real value approach have been developed to address some of these criticisms.

## Conventional treatment of outgoings and voids

### Example 5.5

Value freehold shop premises in a tertiary location let on lease with four years to run at £7,000 a year. The tenant pays the rates and the insurance and undertakes internal repairs. It is worth £12,500 a year net today and rental values for this type of property are continuing to rise in this area.

Assumptions:

1   Current equivalent yields for comparable properties let on IR terms are 8%.

2   The valuation is on in arrears assumptions.

Market valuation:

| | | |
|---|---|---|
| Gross income per year for 4 years | | £7,000 |
| Less landlord's outgoings | | |
| External repairs and decorations[1] | £650 | |
| Management at 5% of £7,000[2] | £350 | £1,000 |
| | | |
| Net income | | £6,000 |
| PV £1 pa for 4 years at 8%[3] | | 3.31      £19,860 |
| Plus reversion in 4 years to[4] | | |
| Net income | | £12,500 |
| PV £1 pa in perp def'd 4 years at 8%[5] | | 9.19      £114,875 |
| Estimated market value | | £134,735 |

Market value (say) £135,000[6]

Notes:

1   Based on office records, etc.

2   Based on fees charged by management department for comparable properties and net of VAT.

3   The assumption here is that there is evidence of equivalent yields on net incomes.

4.   In a normal market valuation the reversion is to market rent as estimated in today's market terms. There is no deduction for outgoings here as the market evidence has provided a net equivalent rent. On IR terms the rent would be greater so that less the annual expenses of £1,000 the net rent would still be £12,500.

5   PV £1 pa in perp. deferred = PV £1 pa in perp. at 8% x PV £1 in 4 years at 8% or PV £1 pa in perp. at 8% less PV £1 pa for 4 years at 8%.

6   This final estimate is rounded as no valuer can truly value to the nearest £; figures of this magnitude would bear rounding to the nearest £1,000.

The idea of reducing the term capitalisation rate by 1% to reflect money security is inappropriate in an equivalent yield valuation. Buyers will not normally differentiate between a property let below market rent and one let at market rent because there is no real difference in risk, provided the sum paid for the investment reflects the difference in current income and any difference in expectation of future rental change and that the security of income in terms of remaining lease length and tenant quality are the same. However, on the latter point, is it reasonable to use the same discount rate for both the capitalisation of, and the four-year deferment of, the reversionary income?

A number of valuers would comment that this rate and the deferment rate should be higher by 1 to 2% to reflect the greater uncertainty of receiving the future increased income. The concept of increasing the deferment rate may seem logical, as 8% could be applied to £12,500 if it was receivable today. However, provided there is no obvious extra risk the equivalent yield approach is supportable. A short-cut DCF approach may be preferred or used as a check.

A comparable conventional layer method valuation is more complex due to the outgoings, but is possible with care, as shown in Example 5.6.

---

### Example 5.6

| | | |
|---|---|---|
| Layer income (net) | £6,000 | |
| PV £1 pa in perp at 8% | 12,5 | £75,000 |
| Marginal income [1] | £6,500 | |
| PV £1 pa in perp def'd 4 years at 8% | 9.19 | £59,735 |
| Estimated market value | | £134,735 |
| Value (say) £135,000 | | |

Note: The reversionary income is £12,500 less the net layer income of £6,000; care is needed here if adopting the layer approach without pausing to consider the impact of the outgoings.

---

The conventional capitalisation approach is perfectly acceptable under normal conditions where property is let on normal terms with regular rent reviews, and where there is evidence of capitalisation rates. In such cases it is most logical to use an equivalent yield approach, whether it be in the format of the term and reversion or the layer method.

Where the income pattern is not normal, there is a strong case for using the DCF approach. DCF directs the valuer to concentrate upon an explicit consideration of the net current and future incomes, and upon the correct rate of interest to use to discount that specified cash flow.

## Allowing for voids

---

### Example 5.7

A property is let at £100,000 a year on FRI terms, i.e. net. The market rent on the same terms is £120,000. The lease is due to end in three years and will take one year to re-let and a rent free period of six months will have to be offered as an incentive. There is evidence of market capitalisation rates of 7% for similar properties with 10 years or longer to run.

The choice is between making an adjustment to the yield to reflect the risk of the void and allowing for a break in income.

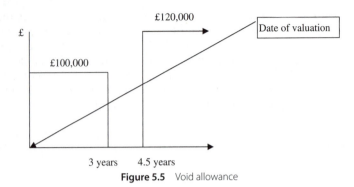

**Figure 5.5**   Void allowance

### Adjusting the yield
Term

| | | |
|---|---|---|
| Current rent | £100,000 | |
| PV £1 pa for 3 years at 7% | 2.6243 | £262,430 |

| | | |
|---|---|---|
| Reversion | | |
| MR | £120,000 | |
| PV £1 pa in perp. defrd. 3 years at 8% | 9.9229 | £1,190,748 |
| Market value | | £1,453,178 |

This adjustment is suggesting that the void risk is now an extra risk to income in perpetuity which would not be the case given the fact that the market expectation is a reasonable one of letting within a year.

**Allowing for a break in income**

| | | |
|---|---|---|
| Term as above | | £262,430 |
| Reversion | £120,000 | |
| PV £1 pa in perp. at 7% | 14.285 | |
| | £1,714,286 | |
| PV £1 in 4.5 years at 7% | 0.737518 | £1,264,317 |
| Market value | | £1,526,747 |

Allowing for a void in this format allows the yield from the market to be retained and for the risk to be allowed for by a specific extension to the period of deferment.

However, as shown in Example 5.8, the only valid approach is to use a DCF because neither of these approaches can be used easily to reflect the costs of UBR, etc whilst the property is vacant.

In multiple let property, such as shopping centres and similar properties, there will be evidence of the normal occupancy rate, for example out of 100 units in a centre 90 might consistently be let. A capitalisation can be undertaken on this basis, using 90% of the rents collected and 90% of the market rent adjusted for the non-recoverable 10% of outgoings.

In some instances it will be found that the market rent is less than the rent being paid, the property is over-rented and can be viewed as a falling annuity (see Figure 5.6).

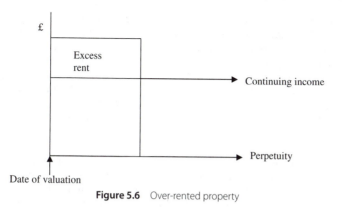

**Figure 5.6**   Over-rented property

## A falling annuity – over-rented property

The issue of falling incomes is not new as it tends to occur when economic slump follows economic boom.

A simple approach is to capitalise the excess income at a higher rate as it is at risk and will cease to exist.

---

### Example 5.8

Value a freehold office building let at £100,000 a year with five years to run. The current market rent is £75,000 a year. The pattern of income is as shown in Figure 5.6.

The valuer's immediate response is 'will rents rise to the contracted level over the next five years?' The key question then is whether this is a rent review and what the terms of the rent review clause are? Or is it a lease renewal? If the latter then the valuer must accept that on current evidence the rent will fall to £75,000. If the former then the valuation problem is partly removed and £100,000 pa could be treated as a perpetual income provided a current capitalisation rate can be derived from market comparables. However, there might be a break clause or it might be upward/downward or it may simply be that the tenant is strong enough to persuade the landlord to accept a reduced rent until such times as market rents rise to £100,000.

Assuming it is a lease renewal and that rents are not expected to rise then a simple solution is for the excess income to be valued at a high capitalisation rate. Here it is assumed that 8% is the normal market yield or capitalisation rate.

| | | |
|---|---|---|
| Continuing income | £75,000 | |
| PV £1 pa in perp at 8% | 12.5 | £ 937,500 |
| Plus income to be lost | £25,000 | |
| PV £1 pa for 5 years at 10% | 3.7908 | £94,770 |
| | | £1,032,270 |

The £25,000 has been capitalised at a higher rate but there is a danger of double counting, as the use of 8% for the continuing income could be implying that the figure of £75,000 is growing and so a part of the term income has been valued twice.

The valuer needs to distinguish the genuine terminating income from that which may represent a temporary shortfall. Thus, a building with a time-limited planning consent may display an income pattern which is going to fall and this reduction will not be recovered until the lower valued use has, through real and inflationary growth over several years, risen to the higher value use. In these cases a hybrid capitalisation may be necessary to reflect not only the fall in income but also the variations in 'risk' attaching to the tranches of income arising from different uses.

### A variable annuity

It is fairly rare to find a completely variable income from property, i.e. one where the income changes from year to year. However, there are two categories which present a degree of annual variability and DCF has to be the preferred approach for both. These are properties let on terms where the rent is indexed to the Retail Price Index (RPI) or alternative indices, and properties let on turnover rents.

The suggested technique is to treat the calculation as a PV calculation, treating each payment as a separate receipt and discounting each to its PV at an appropriate rate.

## DCF approaches

### The short-cut DCF

This method or approach was developed during the property cycle boom years of the late 1970s and early 1980s. The explanation that follows sets out a situation which was typical of that period when investors were buying for capital growth, prices were rising and yields had fallen far below the fixed rate of return available from investing in government securities, such as gilt edge stock. The explanations and examples that follow have not been amended to reflect the very changed position of the post credit crunch years after 2007.

In Example 5.2, no account was taken of the fact that in year five there would be a review of the rent to the market rent. Fixed interest securities are considered to be risk-free; even more so where the value (stock face value) is index linked. So, if at the time of valuation property of a particular type and quality is selling on a 5.5% basis and risk-free stock can be purchased to achieve a return of 10%, then the market for property must be implying an expectation of growth. (The position at the time of the sixth edition is very different in that interest rates have fallen to all time lows, major investors have had to rebalance their portfolios, stock redemption yields have fallen and property market capitalisation rates have risen and remain mainly above redemption yields.) However, when the reverse yield gap becomes pronounced, as it did in the latter part of the twentieth century, the investor in property is buying for growth and this is also the case when initial yields (cap rates) are below investors' target rates, i.e. below the total returns that investors need to compensate for all the risks of investing in property.

Thus, £100 invested in perpetual stock at 10% will be producing £10 per year when the same invested in property is producing 5.5% or £5.50 per year. This represents a loss of £4.50 a year whilst at the same time the investor is accepting the greater risks attached to the purchase of property. If these greater risks warrant a 2% adjustment then the property investor should be looking for an overall return of 12% (10% + 2%). The difference between 5.5% and 12% was at that time known as the 'reverse yield gap', it being the opposite of the normal expectation that investors' returns rise as risks rise. The gap must be made good through growth.

The necessary level of growth can be calculated using DCF techniques, and a number of formulae and tables have been produced to simplify the process. In the above paragraph an annual growth rate of 6.5% might be inferred (12% – 5.5% = 6.5%) but this would not reflect the fixed rent between rent reviews.

The rate of implied growth can be calculated from any one of a number of formulae:

$$K = e - (ASF \times P)$$

Where:

$K$ = the capitalisation rate expressed as a decimal

$e$ = the overall return or equated yield expressed as a decimal

ASF = the annual sinking fund to replace £1 at the overall or equated yield over the review period ($t$)

$P$ = the rental growth over the review period from which the implied rate of rental growth can be calculated.

However, some commentators prefer the formulation:

$$(1+g)^t = \frac{PV£1PA \ perp \ at \ K - PV£1PA \ t \ years \ at \ e}{PV£1PA \ at \ K \ x \ PV \ t \ years \ at \ e}$$

So, given that K = 0.055 (5.5%), $e$ = 0.12 (12%) and $t$ = 5 years, then:

0.055 = 0.12 – (0.1574097 × P)

0.1574097 P = 0.12 – 0.055

0.1574097 P = 0.065

P = 0.065/0.1574097

P = 0.412935 (which multiplied by 100 to convert to a % = 41.29% over five years).

The nature of compound interest was outlined in Chapter 2. Here, there is an implied growth in rent, P, over five years, of 41.29% and therefore:

$$P = (1+g)^t - 1$$

Where:

$g$ = growth per year expressed as a decimal and

$t$ = the rent review period (in this example this is every five years)

1 + P must equal $(1 + g)^t$ to give

1.412935 = $(1 + g)^5$

$1.412935^{0.20}$ = $(1 + g)$

[(or $\sqrt[5]{1.412935}$ = $(1 + g)$]

$1.071579 = (1 + g)$

Therefore $g = (1.071579 - 1) \times 100$

$g\% = 7.1579\%$.

The figure 7.1579% represents the implied annual average rental growth in perpetuity.

The steps in the short-cut DCF method can now be stated:

- Discount the current rent to the next review or lease renewal at the overall rate or equated yield. This is logical as there can be no growth in a rent fixed by legal contract (the lease) for a term of years.
- Multiply the estimated market rent by the amount of £1 for the period to review at the implied rate of rental growth.
- Discount (capitalise) the future market rent in perpetuity at the capitalisation rate (ARY) and discount the capitalised value on reversion for the intervening period at the overall or equated yield.

The issues are:

- What is the ARY (capitalisation rate)?
- What is the overall yield (equated yield)?

### Example 5.9

Revalue the property in Example 5.2 using a short-cut DCF approach on the basis of an equated yield of 12% and an ARY of 5.5%.

| | | | |
|---|---|---|---|
| Current rent | | £120,000 | |
| PV £1 pa for 5 years at 12% (YP) | | 3.6048 | £432,576 |
| Rent in 5 years | £120,000 | | |
| Amount of £1 in 5 years at 7.1579% | 1.412931 | | |
| Implied rent in 5 years | | £169,552 | |
| | | | |
| PV £1 pa in perp at 5.5% | 18.1818 | | |
| PV £1 in 5 years at 12% | 0.5674269 | 10.3168 | £1,749,243 |
| | | | £2,181,819 |

The value is identical to the normal income capitalisation of £2,181,818 (bar £1 due to rounding error) because of the underlying assumptions incorporated in the implied rental growth formula. This could be set out as a cash flow over a number of years if preferred by the client.

The figure of £169,552 does not represent a forecast of the rent to be expected in five years time. It is simply an expression of future rental derived from the market's implications of purchasing risk investments below risk free or target rates. It represents the rental needed in five years time to recoup the loss of return (income) during the next five years in order to achieve an overall return 2% higher than the given risk-free rate.

### Question 1: Implied growth rate calculation

Freehold shops in prime locations on FRI terms, with rent reviews every five years are selling on an initial yield (ARY) of 6%. If investor's target rates are 1% above gilts standing at 8% what is the implied rental growth?

Professor Neil Crosby (University of Reading) developed a real value approach based on the earlier work of Dr Ernest Wood and has become the principal exponent of this technique. The technique is similar to the short-cut DCF method in that it values the current rent at the equated yield. Where it differs is that it retains the reversionary rent at today's market rent but discounts the reversion to capital value at the Inflation Risk Free Yield (IRFY).

When using any of the capitalisation techniques, the valuer needs to assess the nature of the rental agreement as set out in the lease to ascertain, in the light of current market conditions, the extent to which the rental income is inflation proof.

- Where rental income is completely inflation prone – that is where it is fixed with no expectation of change – the equated yield should be used in the discounting process.
- Where rental income is completely inflation proof the IRFY should be used.

This process reflects the fact that between rent reviews the real value of the income may be falling.

The steps in the method follow the DCF method in assessing the implied growth rate; from this the valuer calculates the IRFY from the formula:

$$\frac{e-g}{1+g}$$

In this example:

$$\text{IFRY} = \frac{0.12 - 0.071579}{1.071579} = \frac{0.048421}{1.071579} = 0.04518659 \times 100 = 4.5187\,\%$$

hence the valuation becomes:

|  |  | £120,000 |  |
|---|---|---|---|
| PV £1 pa for 5 years at 12% |  | 3.6048 | £432,576 |
|  |  | £120,000 |  |
| PV £1 pa in perp at 5.5% | 18.1818 |  |  |
| × PV £1 in 5 years at 4.5187% | 0.80173 | 14.577 | £1,749,235 |
|  |  |  | £2,181,811 |

(Small discrepancy of £7 due to rounding of some numbers.)

Both the DCF and real value models overcome some of the key criticisms of conventional valuation when valuing property currently producing less than the market rent.

---

### Example 5.10

Using the facts from Example 5.4, revalue on a modified DCF basis assuming the market rent of £120,000 is based on a normal five-year review pattern and that an equated return of 12% is required.

The implied growth rate is 7.1579% (see p.110)
This implies that the rent in two years time would need to be:
£120,000 × amount of £1 at 7.1579% for 2 years
£120,000 × 1.14828 = £137,793.79, say £137,794
The property can now be re-valued on a contemporary basis using the modified or short-cut DCF method (Baum and Crosby (2008) provide a full critical comparison).

Short-cut DCF

| | | |
|---|---|---|
| First 2 years | £100,000 | |
| PV £1 pa 2 years at 12% | 1.6901 | £169,010 |
| Reversion to | £137,794 | |
| (£120,000 × A £1 | | |
| at 7.1579% for 2 years) | | |
| PV £1 pa in perp at 5.5% | 18.1818 | |
| | £2,505,339 | |
| PV £1 at 12% for 2 years | 0.7972 | £1,997,256 |
| | | £2,166,266 |

The difference in opinion in practice on such a short reversionary property would be less significant as the valuations would probably be rounded to £2,165,000 (Example 5.7) and £2,145,000 respectively. However, it can be seen how the term and reversion produces a nearly acceptable solution by overvaluing the term by £15,620 (£184,630 – £169,010) which to some extent compensates for the undervaluation of the reversion by £36,996 (£1,997,256 – £1,960,260).

The short-cut DCF now reads like a valuer's investment valuation as all contracted income is discounted at a money market rate or target rate (the equated yield) and the reversion is to an expected, albeit implied, rent, not today's market rent.

The importance of this issue is more evident when considering a longer reversionary property.

### Example 5.11

Assuming the same facts as previously but assuming a reversion in 15 years.
Conventional
Term and reversion

| | | |
|---|---|---|
| Current rent | £100,000 | |
| PV £1 pa for 15 years at 7% | 9.1079 | £910,790 |
| Reversion | £120,000 | |
| PV £1 pa in perp at 5.5% | 18.1818 | |
| | £2,181,818 | |
| PV £1 in 15 years at 5.5% | 0.4479 | £977,236 |
| | | £1,888,026 |

A capitalisation rate of 7% has been used for the term rather than 5.5% in order to reflect the difference between an investment with a long reversion and one with a short reversion. This adjustment is subjective. The reversion is still at 5.5% because it should be given the same value as the income of £120,000 in previous examples for the period from year 15 into perpetuity.

However:
£120,000 × Amount of £1 at 7.1579% for 15 years is
£120,000 × 2.821 = £338,489
Short-cut DCF

| | | |
|---|---|---|
| Current rent | £100,000 | |
| PV £1 pa for 15 years at 12% | 6.8109 | £681,090 |
| Reversion | £338,489 | |
| PV £1 pa in perp at 5.5% | 18.1818 | |
| | £6,154,336 | |
| PV £1 in 15 years at 12% | 0.1827 | £1,124,397 |
| | | £1,805,480 |

The modified DCF assesses the fixed lease income at the equated yield or opportunity cost, the reversion is to the implied rent in 15 years time and the value is deferred at the equated yield. It is extremely difficult in the absence of true comparables to arrive at a correct equivalent yield subjectively. Intuition appears to be the favoured approach of a number of valuers, with some using money market rates for discounting the fixed term in an effort to overcome the weaknesses of the conventional valuation method.

Currently, though, it is not possible to say that a short-cut DCF approach must be the preferred approach. It may be the more rational approach, in that it is possible to argue that investors should be indifferent between the short and long reversionary properties if they are expected to produce the same equated yield. However, this is difficult to support if the market is substantially discounting long reversions through the subjective approach of their investment surveyors; implying inconsistency over the choice of equated yields. The short-cut DCF can be proved by using a full projected cash flow over, say, 100 years, discounted at 12% as in the table below.

DCF to 100 years allowing for a rental growth at 7.1579% discounted at an equated yield of 12% and providing for a reversion to future capital value in year 100. This would bring the value figure more into line with the modified DCF.

| Period in years | Rent x A £1 pa at 7.1579% | PV £1 pa at 12% | PV at 12% | PV |
|---|---|---|---|---|
| 0-15 | 100,000 | 6.8109 | 1.0 | 681,090 |
| 16-20 | 338,489 | 3.60477 | 0.18269 | 222,913 |
| 21-25 | 478,262 | 3.60477 | 0.10367 | 178,729 |
| 26-30 | 675,750 | 3.60477 | 0.05882 | 143,281 |
| 31-35 | 954,788 | 3.60477 | 0.03337 | 114,853 |
| 36-40 | 1,349,049 | 3.60477 | 0.01894 | 92,105 |
| 41-45 | 1,906,112 | 3.60477 | 0.01074 | 73,796 |
| 46-50 | 2,693,203 | 3.60477 | 0.00609 | 59,124 |
| 51-55 | 3,805,307 | 3.60477 | 0.00346 | 47,462 |

(Continued)

| 56-60 | 5,376,633 | 3.60477 | 0.00196 | 37,988 |
|---|---|---|---|---|
| 61-65 | 7,596,806 | 3.60477 | 0.00111 | 30,397 |
| 66-70 | 10,733,755 | 3.60477 | 0.00063 | 24,376 |
| 71-75 | 15,166,045 | 3.60477 | 0.00036 | 19,681 |
| 76-80 | 21,428,560 | 3.60477 | 0.00020 | 15,449 |
| 81-85 | 30,277,055 | 3.60477 | 0.00011 | 12,006 |
| 86-90 | 42,779,360 | 3.60477 | 0.00006 | 9,253 |
| 91-95 | 60,444,240 | 3.60477 | 0.000037 | 8,062 |
| 96-100 | 85,403,480 | 3.60477 | 0.000021 | 6,465 |
| 100-perp | 2,193,984,364 | | 0.00001 | 21,940 |
| | | | | 1,798,970 |

In practice, such contemporary methods seem to be rejected in favour of the intuitive market methods. This implies that although valuers do not make the market they do in some instances have a strong influence. Current market practice favours the simple equivalent or same yield approach.

This creates the probability that investors will conclude that some valuers are overvaluing or undervaluing properties and will sell or buy in the market at market prices to take advantage of such market imperfections.

Throughout this section it must be remembered that the growth rates used are implied and that the figures derived in no way predict the future. They merely provide the valuer with an additional tool with which to examine and, as will be seen later, to analyse the market. Such implied rents must be critically examined against the reality of the marketplace and realistic economic projections.

The position in 2011 is that the valuation profession is using DCF to a much greater extent, both to value property and to assess investment value or worth. The need has arisen from the problems of the recession where the short term income profile of many properties does not match that associated with the conventional term and reversion and layer methods. Using DCF requires the valuer to be market specific about the target or hurdle rates being adopted and about the expected rents over the short term. Example 5.12 illustrates how a DCF solution might be developed.

### Example 5.12

Following a series of bank mergers and acquisitions, a retail bank now has five units within 100 metres of each other in the Anytown centre. The bank has announced that through rationalisation two will close on expiry of their leases. You are required, for secured lending purposes, to value the freehold interest in one of the units which has been identified for closure. It is currently let on FRI terms on a lease due to terminate in two years time. The current rent is £100,000. The market rent was £120,000 but is now only achievable as a headline rent. Recent lettings of vacant units arising from retail closures following administration have generally taken a year to secure a tenant and have then let at headline rents with one year rent free, plus pre-let works other than unit fronts and tenant fittings. It is expected that the same incentives will apply to this unit.

The effect on rent roll will be as shown in the cash flow. Management, insurance, repairs and empty property rates (UBR) are best estimates. Repairs in year three are the pre-let costs to be borne by the landlord and ongoing management is expected in year three. No allowance has been made for agents' letting fees but these could also be included.

The estimates in this example are assumed to have been derived from economic analysis and suggest that market rents will regain lost ground over the next four years and will then begin to rise at the long term rate of inflation rate of 2.5%. The discount or hurdle rate is based on gilts (risk free rate) at 4.5%, property risk 3.0%, plus tenant risk 0.5%, lease risk (likely to be short) 0.5%, building risk 0.5%, location 1%, giving a discount rate of 10%.

Market value in year 10 is based on the market rent plus inflation for five years at 2.5%. Market yields for this type of town centre retail property are currently expected to stabilise at their pre-credit crunch level, which in this case is taken as 6%.

Valuing on a term and reversion basis at an equivalent yield is unlikely to be sufficiently explicit and will require the valuer to base the valuation on sentiment, feel, intuition, experience. For secure lending purposes it is possibly preferable to use a DCF along the lines suggested below. This may still be insufficient as it assumes a degree of certainty as to future events. This would suggest the use of what if? scenarios; inclusion of probabilities, or some form of sensitivity analysis. Methods are not prescribed by The Red Book but for secured lending The Red Book processes must be followed. The calculations are based on an Excel spreadsheet which may guarantee some accuracy in the arithmetic but the assumptions must all be supportable with market evidence and clear reasoning.

| Discount rate | 10% |
|---|---|
| Terminal Cap Rate | 6% |

| End Year | Income | Mng | Ins | Reps | UBR | Net | PV at 10% | PV |
|---|---|---|---|---|---|---|---|---|
| 1 | 100,000 | 0 | 0 | 0 | 0 | 100,000 | 0.90909 | 90,909 |
| 2 | 100,000 | 0 | 0 | 0 | 0 | 100,000 | 0.82645 | 82,645 |
| 3 | 0 | 3,000 | 8,000 | 5,000 | 42,000 | −58,000 | 0.75131 | − 43,576 |
| 4 | 0 | 3,000 | 8,000 | 5,000 | 42,000 | −58,000 | 0.68301 | − 39,615 |
| 5 | 120,000 | 3,000 | 0 | 10,000 | 0 | 107,000 | 0.62092 | 66,439 |
| 6 | 120,000 | 0 | 0 | 0 | 0 | 120,000 | 0.56447 | 67,737 |
| 7 | 120,000 | 0 | 0 | 0 | 0 | 36,000 | 0.51316 | 18,474 |
| 8 | 120,000 | 0 | 0 | 0 | 0 | 36,000 | 0.46651 | 16,794 |
| 9 | 120,000 | 0 | 0 | 0 | 0 | 36,000 | 0.42410 | 15,268 |
| 10 | 120,000 | 0 | 0 | 0 | 0 | 40,000 | 0.38554 | 15,422 |
| 10 | 2,262,816 | 0 | 0 | 0 | 0 | 2,262,816 | 0.38554 | 872,414 |
| Present Value | | | | | | | | **1,162,909** |

## DCF and the over-rented property

The 1979 edition of this book suggested that a DCF approach would be the preferred solution to the falling income problem. Crosby, Goodchild and others have developed this approach to the valuation of the over-rented property.

The steps in the solution are as follows:

1. Assess the implied rental growth rate from comparable investment properties let at their market rent.
2. Calculate the point in time when current market rent will exceed the current contracted income.
3. Capitalise the contracted rent up to the cross-over point or first review thereafter at the equated yield.
4. Capitalise the implied market rent at the equivalent yield (ARY) from the date taken in step 3 in perpetuity discounted for the period of waiting at the equated yield.

A solution can be found to Example 5.8 using the Crosby approach:

1. ARY from comparable properties are 8%.
2. Equated yield is 12%.
3. Implied rate of rental growth given $e = 0.12$ and $k = 0.08$ is 4.63%.
4. 7,500 x Amount of £1 in '$n$' years at 4.63% = £10,000

$$7,500 \times 1.334 = £10,000$$
$$(1 + 0.0463)^n = \sqrt[n]{1.334}$$
$$n = \sqrt[n]{1.334}$$
$$n = > 6 < 7$$
say 6.5 years

5. In this example, given an upward only rent review in five years, the rent will remain at £10,000 at the first review. The valuer must exercise judgment to decide the most probable date for a review to occur. In this case the valuation might be:

| | | | |
|---|---|---|---|
| Rent | | £10,000 | |
| PV £1 pa 7 years at 12% | | 4.5638 | £45,638 |
| Rent | | £10,000 | |
| PV £1 pa in perp at 8% | 12.5 | | |
| PV £1 in 7 years at 12% | 0.45235 | 5.6543 | £56,543 |
| | | | £102,181 |

If the lease was due to terminate after five years then the implied rent at that time could be calculated at the implied growth rate of 4.63%.

Then:

£7,500 × amount of £1 in 5 years at 4.63% = Implied rent in year 5.

The contracted rent would be valued at the equated yield for five years with a reversion to the lower rent in year 5. Here, there are practical difficulties for the valuer, as most landlords would wish to retain the higher rent by payment of a reverse premium in order to underpin future rent reviews, but valuers must always be conservative in their assumptions in a market valuation.

Whilst theoretical solutions to this problem can be found using implied growth rates and DCF techniques, these may not truly reflect behaviour in the marketplace and the valuer must not let the mathematics override market judgments. In particular, future rents that have been calculated using implied growth rates over the short term must not be used as a substitute for market analysis. The latter may well suggest minimal change in rental value over the short term.

The problem of falling incomes becomes more acute in the secondary and tertiary markets; that is, when considering over-rented properties in non-prime locations, or with non-prime tenants. Here, market capitalisation rates will generally be much higher than gilt rates and the valuer will have to use his or her best judgment in circumstances where the calculation of implied growth is unrealistic and impracticable.

Baum and Crosby (1998) suggest that:

> for a property over-let to a very secure covenant under a lease with more than 15 years unexpired (on upwards only rent reviews), the term yield could be determined in comparison with a fixed income gilt plus a small risk margin for property illiquidity but without the traditional risks of property such as uncertainty for future income flow.

This is an interesting 1990s observation which flows logically from an earlier observation in this book that:

> valuers in the nineteenth century were primarily dealing with rack rented properties, largely agricultural, at a time when inflation was relatively unknown. A realistic and direct comparison could be made between alternative investments such as undated securities (gilts) issued by government. Property being more risky was valued at a percentage point or so above the secure government stock.

The market in the early 2000s again suggested that good quality property was selling at or just above the gilt rate whether or not it was over-rented, provided the lease term was reasonably long and the tenant was secure.

The position in the latter part of the 2000s changed again, post the credit crunch. Property became over-rented and initial yields, security of the tenant covenant and length of the unexpired term became critical factors in determining yields.

In some respects history is repeating itself. It is again possible to argue that the original 'term and reversion' valuation may be appropriate in some circumstances. Thus, an over-rented property may show little prospect of income change to the owner until a future lease renewal, at which point best estimates or implied growth may suggest a market rent above the current contracted rent. The future rent may appear to be less secure, as it will depend on a new letting or lease renewal and at such a date in the future that further rental growth may be unlikely due to depreciation and economic obsolescence. A market conventional approach might be to capitalise the term at, say, 1% above comparable gilts reflecting current income security of the tenant in occupation, and the reversion at 2% above gilts to reflect the greater risk attached to the reversion including the risk of a letting void.

The reader is reminded that, whilst detailing income valuation methods and the links between conventional and DCF techniques, the real task is the identification of the risks associated with the property to be valued, which requires research into future expectations (forecasting) as well as observation of the current market and market transactions.

The use of non-conventional methods is inappropriate where there is ample evidence of recent transactions and where the capitalisation is of a property let at the current market rent on current lease terms. Otherwise, a DCF approach is likely to be required to show explicitly the assumptions that have been made and why.

### Question 2: Value a freehold warehouse

A comparable property on the same business park let at its market rent of £42,500 has just been sold for £472,000. The subject property is similar in size but was let on a reduced rent of £30,000 three years ago. There is a rent review in two years time. Assess the ARY and value the warehouse. Demonstrate the use of modified/short-cut DCF and real value approaches on the basis of an equated yield requirement of 12%.

## Advance or arrears

It has always been the custom to assume annual 'in arrears' income flows and annual rates for capitalisation purposes. There is, however, a growing tendency, as property is let on 'in advance' terms, to value on such a basis. In this respect it is essential to determine the valuation date at the commencement, as in many cases this will fall between rental payments. Technically, this is still an 'in advance' valuation as the assumption of market value will generally imply an apportionment of rents received 'in advance' as at the date of completion of a sale, and for valuation purposes the valuation date may be treated as a sale completion date.

Second, one should not lose sight of the fact that 'in advance' means one payment due immediately, i.e. its present value is the sum due, and thereafter the same sum is due for $n$ periods in arrears.

The crucial point is the relationship between the income and the discount rate. The latter must be the correct 'effective' rate for the particular income pattern. For preference this should be derived from market analysis of comparable 'quarterly', 'yearly', 'in advance' or 'in arrears' transactions as investors may not in fact be prepared to accept the same effective rate.

If a property valuation is analysed on a precise basis, allowing both for the correct apportionment of rent as at the date of valuation and for the correct timing of future rents, then the rate of return will be higher than the rate per cent adopted in the valuation on an 'annual in arrears' basis. This realisation has encouraged some valuers to switch to 'quarterly in advance' valuations using published tables or programmed calculators and computers. Where this is done, valuers must be sure that they adopt the proper market relationship between income patterns and yields of comparables in their valuation work. For example, the PV £1 pa in perpetuity annual in arrears at 10% is 10 but using 10% and substituting in the quarterly in advance formula produces a multiplier of 10.6176. Hey presto, values have gone up! However, this is not the case; values are not a function of the maths but of the market. Switching to quarterly demands a switch in analysis, and for valuations to be based on the corrected yields which, substituted in the quarterly formula, produce the same value as when using annual in arrears assumptions; hence the strong reluctance to simply change for the sake of change.

When the whole market relates to 'quarterly in advance' lettings and all valuers are analysing yields on a precise basis, then the market yields adopted by valuers for capitalisation work will be correct for 'in advance' valuations. Until then the valuer needs to be fully aware of the basis of the quoted market yield before transposing it to an 'in advance' valuation.

The reader should note that as the assessment of rental value, outgoings and capitalisation rates are all opinions, switching to 'in advance' will not in itself achieve a better opinion of value. Attempts at such arithmetic accuracy may be spurious.

Save where stated, the 'annual in arrears' assumption has been used in all examples. This assumption is still commonplace in valuation practice, but accurate investment advice requires that estimation of the exact timing of income receipts is necessary in order to assess yields accurately. The use of spreadsheets enables the valuer to incorporate an accurate calendar within any valuation or analysis programme.

On examination, it will be noted that certain property investments resemble certain other forms of investment and they can be distinguished by the future pattern of returns. Thus, property let on long lease without review could be compared to an irredeemable stock.

Owner-occupied freehold commercial properties are comparable to equity shares, whereas freehold properties let on a short lease or with regular rent reviews (whilst in a sense resembling equity investments) must also reflect the stepped income pattern. Short fixed income leasehold interests are comparable to any fixed term investment such as an annuity.

Recognition of these relationships is essential if the valuer is to make correct adjustments to the capitalisation rates to be used in a valuation where the income pattern produced by the property is out of line with current market evidence.

## Analysing sale prices to find the equivalent yield

The equivalent yield is the IRR assuming no growth in rent. The simplest explanation is that this analysis is the reverse of a valuation. In Example 5.4, a property was let on FRI terms with two years to run at £100,000. The market rent was £120,000. A valuation at 5.5% produced a value of £2,144,890 say £2,145,000. If at a later date the property was sold for £1,750,000 then that is the market price that needs to be analysed to establish the equivalent yield. All the information is known except the yield that produces a value of £1,750,000. The simplest and most accurate solution is Goal Seek in Excel, but traditionally students will be taught to derive the yield from trial and error.

- Step 1. Set out the valuation in conventional term and reversion or layer form.
- Step 2. Consider the relationship between the market rent and the sale price to assess a ceiling yield.
- Step 3. Substitute a yield close to the ceiling yield in the valuation and test to see if this produces a value of £1,750,000.
- Step 4. Repeat until a close solution is found and interpolate.

Step 1:

| Term | | |
|---|---|---|
| Current rent | £100,000 | |
| PV £1 pa for 2 years at x% | 0.00 | £000,000 |
| Reversion | | |
| Market rent | £120,000 | |
| PV £1 pa in perp defd 2 years at x% | 0.00 | £000,000 |
| Market value (sale price) | | £1,750,000 |

Step 2:
$$(120,000 / 1,750,000) \times 100 = 6.86\%$$

This would be the initial yield (ARY) if let today at £120,000 and so the equivalent yield must be below this figure as the current rent is only £100,000.

Step 3: Try at 6.75%. At this equivalent yield the term would be valued at £181,430 and the reversion at £1,560,000, giving a total of £1,741,430.

Step 4: Repetition at this level without using Goal Seek would be tedious and very time consuming. Clearly, as the value is just below £1,750,000, the actual yield must be fractionally under 6.75%. Given the fact that sale prices are negotiated then it might be reasonable to assume that investors are willing to buy this quality of freehold property on the basis of equivalent yields (IRRs) of 6.75%.

Goal Seek produces a figure of 6.72%

# A final adjustment

The last steps in any freehold valuation are as follows:

- Reflect on the final figure and review the value sum in the light of market knowledge, some valuers might call this the 'feel right factor'.

- Adjust the final figure to indicate that valuation is not a precise science and any value given to the nearest £ would suggest a high degree of accuracy; the adjustment could be to round to the nearest appropriate round number given the magnitude of the valuation, e.g. if the calculation comes to £10,565,212 the valuer might round to £10,550,000. There is no rule on this and only practice and experience can guide the valuer.

- Adjust the figure before or after rounding to reflect that any acquisition of property will incur considerable on costs to cover SDLT, solicitors' fees and associated charges, building surveyor and valuation surveyor fees; some of which attract VAT at 17.5%. SDLT is at 4% on sales above £500,000 and an overall adjustment of 5.75% is made on valuations at this level, or of 4.75% if the valuation is between £250,000 and £500,000. The need for this adjustment is because a sale at, say, £1,000,000 for an income of £100,000 on a 10% capitalisation basis becomes a purchase at a total cost of £1,057,500 which provides a return of 9.46%, which is not what the valuation implied. A division of the figure by 1.0575 produces the sum which, if it becomes the contract price, will provide the investor with a 10% return on total cost. £1,000,000/1.0575 = £945,626 and £945,626 + 5.75% = £1,000,000. This adjustment of 5.75% is often expressed as 5.7625% with VAT rising to 20%. Spurious sophistication may adjust this to 5.80% (4.0% SDLT fees of 1.5% + VAT at 20% = 4.0%+(1.5%+0.3%)=5.8%). As an observation, many investors use retained surveyors and solicitors, or in house surveyors and solicitors and the per property cost is possibly less than 0.25%.

These last steps are typical, but given that valuation is both an art and a science some valuers will go no further than refining their opinion based on experience of the marketplace – the 'feel right factor'.

# Sub-markets

RICS requires its members to accept instructions to value only if they have the expertise within the specified market to be able to provide professional advice. In this respect it is worth noting, that whilst the general principles of freehold investment valuation methodology will apply to all parts of the market, expertise or market experience is often very specialist and may relate to a relatively small sub-market or sector.

The following list is indicative of some of the sub-markets.

- RESIDENTIAL
- COMMERCIAL
  - RETAIL
    - Single shops
    - Shopping centres
    - Out-of-town centres
    - Retail parks
    - Factory outlet parks
    - Garden centres
  - OFFICES
    - City centre
    - Business park
- INDUSTRIAL
  - Heavy industry
  - Light industry
  - Starter factories
- AGRICULTURAL
  - Farms
  - Estates
  - Market gardens
- LEISURE AND PLEASURE
  - Hotels
  - Licensed premises
  - Leisure centres
  - Cinemas
  - Adventure parks
- HEALTH
  - Private hospitals
  - Health clinics
  - Medical practice centres
  - Dental practice centres
  - Physio centres, etc.

Within these and other sectors there are divisions between freehold and leasehold; owner-occupied and tenanted; market rented; under-rented; over-rented. Throughout, the valuer needs to establish risk and an appropriate capitalisation rate for conventional valuation or target/hurdle rate for DCF.

# Summary

- All freehold income valuations are based on the principle of discounting future net benefits to their present value.
- Two approaches are used:
  - Conventional capitalisation at the ARY or cap rate or equivalent yield
  - DCF using an equated yield or target/hurdle rate.
- Conventional approaches are used where there is good market evidence of capitalisation rates.
- DCF is used where it is desirable to express implied elements of future changes or lack of changes such as rental growth explicitly. DCF is used for:
  - Valuations, and for
  - Assessment of investment value or worth.
- Assessment of growth rates can be based on implied rental growth or on observable market factors such as rent inflation. Assessment of implied growth must not be confused with the more significant projections based on econometric models.

# Spreadsheet user

To assist in the analysis of property sales and in the valuation of a typical single property investment, the valuer can develop a simple Excel spreadsheet. In practice the unique nature of each valuation may require a property-specific spreadsheet. Many clients, particularly for secured lending, now specify the use of one or other of the standard software packages such as Argus or Kel.

## Project 1

The spreadsheet reprinted here is an illustration of the learning tools that readers, in particular students, can develop to facilitate the calculations required to complete typical valuation tasks. This spreadsheet is designed to calculate an equivalent yield using Goal Seek in Excel, or trial and error and to value using the term and reversion method using an equivalent yield. The first section contains the data required for both calculations. Cells are completed for the data headings and left blank for the subject property data details. Typical costs have been included. The calculations access the cell data as part of the equations set up for the blank cells.

### Term and reversion

For a valuation, complete all data boxes other than the contract price cell. Scrolling down completes the calculations as per the equations created in the PV £1 pa and present value cells. The term and reversion are added together and entered against the market value.

## Equivalent Yield Analysis on Contract Price or Term and Reversion Valuation

| Data | Details | | Cost | |
|---|---|---|---|---|
| Current net of outgoings rent | £500,000 | | | |
| Market rental value net of outgoings | £750,000 | | Stamp Duty | 4% |
| Period to rent review or to lease renewal (e.g. 3.25 years) | 2 | | Legals | 1.0% |
| Trial Rate % | 8.00% | | Suveyor et al | 0.5% |
| Contract Price | £0 | | Misc      VAT | 17.50% |
| | | | Total | 5.7625% |
| **TERM** | | | | |
| Current net rent | £500,000.00 | | | |
| YP for          2 years at   8.00% | 1.7833 | | | |
| Term value | | £891,632.37 | | |
| **REVERSION** | | | | |
| Market rental value | £750,000 | | | |
| YP perp. deferred      2 years at      8.00% | 10.71673525 | £8,037,551.44 | | |
| **Market Value** | | **£8,929,183.8** | | |
| **Less Contract Price** | | **£0.00** | | |
| **Net Present Value** | | **£8,929,183.81** | | |
| **Net Value** | | **£8,442,674.68** | | |
| **Initial Yield on Market value as a %** | | **5.60** | | |
| **Yield on Reversion on Market Value as a %** | | **8.40** | | |

It is regular practice to reduce the market value to reflect typical acquisition costs. This produces a sum which, when costs are added back, brings the figure back to the sum which shows the same equivalent yield used in the valuation. If this is not done and costs are added to the market value sum then the IRR will no longer be the same as the equivalent yield used in the valuation.

## Equivalent yield analysis

If a freehold sale price is known together with the net incomes and unexpired term then the spreadsheet can be completed with all data, including the sale price but not the yield. The spreadsheet user can then try different yields until by trial and error a value is found equivalent to the sale price. This is tedious and so Goal Seek in Excel should be used.

## Other yields

The final cells present the initial yield (year one rent as a return on value) and the yield on reversion. These are often quoted by agents on investment marketing literature.

Project 1 is to develop a similar spreadsheet for your own use.

# Project 2

Creating spreadsheets for the various conventional and contemporary freehold valuation methods is a useful learning tool. Only if the method has been understood will readers and students be able to create their own spreadsheets. For this reason we do not provide the answers but hope learners will find their own solutions and in so doing improve on our suggestions.

Project 2 is about developing spreadsheets for:

- the layer method
- for valuations with outgoings
- valuations on a quarterly in advance basis
- extending the spreadsheet to produce market rent from rental data
- implied rental growth
- contemporary valuations based on DCF with growth.

As spreadsheets do exactly what the user asks them to do, they need to be tested to ensure that they are correct. To begin with readers are advised to adhere to annual in arrears assumptions and to check results against those prepared using pen, paper and present value equations.

# Chapter 6

## The Income Approach: Leaseholds

## Introduction

The problem of valuing leasehold interests in property has been troubling valuers for many years. In most countries the debate is now closed. However, in the United Kingdom and in those countries where valuation methodology has followed UK practice, there is disagreement between those who regard the 'dual rate' (sinking fund) or reinvestment approach to be correct and those who would seek to analyse a leasehold sale or valuation as they would any other investment requiring analysis or valuation, namely to apply single rate principles.

The view of the authors is that the single rate approach is the only objective method of analysis of leasehold sale prices, and hence the only objective leasehold valuation approach as it mirrors the behaviour of buyers. Single rate is illustrated in this chapter, together with the preferred discounted cash flow (DCF) approach. For continuity, the previous debate on dual rate can now be found in Appendix A.

Empirical data collected as part of the Royal Institution of Chartered Surveyors (RICS) research into Property Valuation Methods (1986) indicated that a variety of methods were in use in the marketplace, with a significant 50% using Years' Purchase (YP) dual rate unadjusted for tax. A more recent, albeit restricted, survey undertaken by A. Farrar for his 2009 Master's dissertation at Sheffield Hallam University noted that 69% of valuers used single rate, 11% dual rate, 17% dual rate adjusted for tax and 3% used another approach. In addition, he found that single rate was the predominant method for leases with less than 10 years and over 60 years to run, whilst dual rate was sometimes used for leases over 10 years.

The following key facts should be noted:

1. A leasehold interest has a finite duration.
2. In investors' terms, a lease can only have a market value if it produces an income and if the lease is assignable. The leasehold income is the profit rent that is the rent actually received, or potentially receivable, less the rent paid to the superior landlord and adjusted for any irrecoverable operating costs or outgoings.

3. The net of tax return for taxpayers will be much lower than expected because the capital cost of acquiring a finite investment has to be recovered out of the taxed income.

4. The mathematical formulation of the PV £1 pa, single rate, is:

$$\left(\frac{(1-PV)}{i}\right) \ or \ \left(\frac{1}{i+ASF}\right)$$

The first provides for interest (return on money) on the diminishing capital cost over the life of the investment, which is the normal concept for investments, loans and mortgages; the second provides for interest on initial capital throughout the life of the investment at the capitalisation rate. The use of either equation where $i\%$ is held constant throughout will produce the same multiplier and hence the same present value. That acquisition at that present value will, therefore, secure a return on money invested and a return of money invested over the life of the investment (see Chapter 2).

The crux of the leasehold valuation problem is how the buyer of a leasehold interest should or does allow for capital recovery. Jon Robinson (1989) in *Property Valuation and Investment Analysis: A Cash Flow Approach* provides a very full summary of the confusions that have emerged over the implicit reinvestment assumptions within the internal rate of return. The book includes the following quotation from the work of Merrett and Sykes (1973) implied rental growth:

> Yield, or discounted cash flow return, is correctly defined as the rate of return on the capital outstanding in a project during every year of its life. The interpretation no more assumes that recovered funds are reinvested at the project's rate of return than the statement that a bank is receiving an 8% rate of interest on an overdraft implies that the bank is reinvesting the overdraft repayments at 8% (Merrett and Sykes (1973) p.130 in Robinson (1989) p.83).

In other words, the argument that buyers of leasehold investments need to reinvest to recover capital is inappropriate because that simply is not a proper interpretation of the yield or internal rate of return.

Leasehold interests are valued by valuers and priced by investment agents. Valuations may also be needed for negotiations between landlord and tenants over surrender values, surrender and renewal of leases and in the case of marriage values (see Chapter 8).

The terms and conditions of leases vary considerably. They can be for any term up to 999 years, and with or without rent reviews or break clauses. They can be on full repairing, non-repairing or internal repairing terms and subject to a wide range of restrictions on use, alienation (assignment and sub-letting). The latter is value critical, because unless the tenant can assign the lease then it cannot be transferred and if it cannot be transferred then it cannot have a market value.

The market can be divided between occupational leases and investment leases. The market value of an occupation lease will depend upon the perceived value to an incoming tenant (assignee) of the occupation and financial benefits attaching to the specific lease compared to those attaching to a new lease at market rent, i.e. what is the vacant possession value of an existing lease of a specific property at a given date? These need to be distinguished from the true leasehold investment.

## Occupation leases

Occupation leases may have an assignment value if the rent payable is less than the market rent. This is a question of assessing what capital sum the assignee would be willing to pay the assignor. These leases are often very short, i.e. there may only be one or two years to the next review to market rent or to the lease termination date. Tenant A simply no longer wants the space; and would like tenant B to pay the costs of the assignment and, what in the market is known as, a premium. On a very short unexpired term, tenant A, or their agent, might simply multiply the difference between the market rent and the rent that has to be paid (the profit rent) by the unexpired term, e.g. if the profit rent is £5,000 (market rent £20,000 less rent payable £15,000) and the unexpired term is two years then a price or premium of £10,000 might be asked for the lease. The point that tenant A is making is that for the next two years the assignee will be paying £15,000 for space which should be costing £20,000, so 'pay me what you are saving'. For a longer unexpired term it might be appropriate to discount the profit rent at a typical cost of borrowing rate.

### Example 6.1

A occupies retail space which is surplus to their requirements and wishes to assign their lease. The lease is assignable with the landlord's consent which cannot be unreasonably withheld. The lease has eight years to run without review and the profit rent (rent saving) is £5,000. Banks are currently lending to this quality of tenant at 8%. What price should be put on this lease in the agent's sale details?

| End of year | Profit rent | PV £1 at 8% | Present value |
|---|---|---|---|
| 1 | £5.000 | 0.9259 | £4,629.50 |
| 2 | £5,000 | 0.8573 | £4,286.57 |
| 3 | £5,000 | 0.7938 | £3,969.05 |
| 4 | £5,000 | 0.7350 | £3,675.05 |
| 5 | £5,000 | 0.6806 | £3,402.82 |
| 6 | £5,000 | 0.6302 | £3,150.76 |
| 7 | £5,000 | 0.5835 | £2,917.37 |
| 8 | £5,000 | 0.5403 | £2,701.27 |
| Total | | | £28,732.62 |

Alternatively, as PV £1 pa is the sum of the present values, the £5,000 can be multiplied by the PV £1 pa for eight years at 8% which is £5,000 x 5.7466=£28,733 (discrepancies are due to working to four decimal places only).

The point is that payment of this sum may need to be funded with borrowed money hence the use of a money market borrowing rate. This is a zero position for an assignee who may discount at 10% for the purpose of negotiations. This is only an investment in an accounting sense and the impact on business profit and the new tenant's tax position will need to be discussed by tenant B with their accountants.

Key money may be asked if there is no profit rent. Assignment lease pricing and key money are market sensitive, and will be very dependent on supply and demand at the time of pricing. Excess demand stimulates the level of premiums and key money; excess supply depresses both and may even create a market operating at negative premiums, meaning that a tenant seeking to assign may have to pay an incoming tenant to take on the liability of an existing lease.

The income approach is concerned with the market value and pricing of investments in property. There may be similarities between techniques in the occupier and investment markets, but this chapter is primarily concerned with leaseholds which are bought and sold as investment.

## Investment leases

The following leasehold investments were sold in 2010 by Cushman and Wakefield at one of their auction sales. They give an indication of the problems associated with finding a valuation approach that would fit all leasehold situations.

> Long Leasehold for a term of 150 years from 23rd July 2009 at a current rent of £46,200 per annum with a rent review in 2014 and 5 yearly thereafter to 27.5% of open market rent.... (current gross rent given as £217,622.23 per annum).
>
> Long Leasehold for a term expiring on 15th December 2123 at a fixed rent of £1 per annum (current gross rent given as £188,698 per annum).

The second of these is almost the same as a freehold bar £1, with over 100 years to run. The first has a current profit rent of £171,423.23 but the problem is whether one could assume this sum in perpetuity given the rent reviews to 27.5% of open market rent. In addition there are over 30 units, some on one year licenses and some on short leases and the rent review is to open market rent rather than to a percentage of occupation rents; a clear case of needing to read the legal documents with great care. What is clear is that even if these leases, were for comparable properties in the same location and displaying all the same risk characteristics, other than the lease risk, it would probably be impossible to use the same capitalisation rate due to the nature of the two leases. Investment leases tend to fall into one or other of the following types.

## Medium to long-term leaseholds at a fixed head rent

Where the head lease rent is fixed for the whole unexpired term, but the occupation leases contain rent reviews, then the investment is a growth investment, i.e. it is inflation proof. As such it will be comparable to a freehold but the critical factor will be the length of the unexpired term. Over 60 years is generally considered very comparable to a freehold because the element of income that represents capital recovery is very small. For example, the PV £1 pa for 60 years at 8% is 12.3766 and in perpetuity at 8% it is 12.5; a difference of 0.1234. The capitalisation rate might therefore be 1% to 2% higher than an equivalent freehold rate in a simple capitalisation exercise.

Where the head rent is subject to review, either in line with any sub-lease rent reviews, e.g. at the same time or less frequently, geared directly to the sub-lease rents, e.g. simultaneous review to 5% of occupation leases or otherwise, e.g. head rent reviewed to a modern ground rent, then the problem of the gearing must be noted, and a simple profit rent multiplied by the PV £1 pa approach may not be sufficient to reflect the gearing. A split capitalisation may need to be used as a check, e.g. capitalise the head rent and the sub-lease rent at different rates to reflect the income differences. The more complex problems may well need to be solved using growth explicit DCF methods.

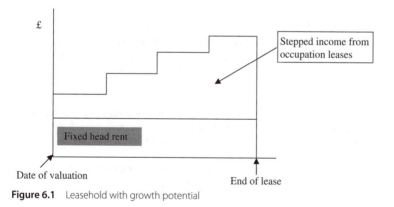

**Figure 6.1** Leasehold with growth potential

## Fixed profit rent

Where the head lessee, or predecessor in title, has sub-let the property for the full unexpired term of the head lease without rent review, the profit rent is fixed. The valuation must reflect this fact and the profit rent must be capitalised at an appropriate cost of capital rate. This effectively allows for the depreciating worth of each year's income in terms of purchasing power. A single rate approach is frequently the most effective method. The yield used will be derived from say the redemption yield on gilt edged securities and then adjusted to reflect the added risks of the leasehold in terms of its covenants and liabilities.

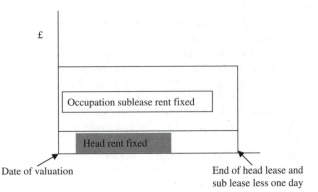

**Figure 6.2**    Fixed income leasehold

# Single rate valuation of leaseholds

All valuations should be based on the analysis of and interpretation of the market. So the first step has to be the analysis of sales of comparable leasehold investments.

### Example 6.2

A leasehold investment has just been sold for £294,500. The lease is of retail premises in a popular shopping street. The lease has 20 years to run and the property has been sub-let for the balance of the lease without rent review. The net profit rent has been assessed at £30,000.

   Trial and error analysis is simply the reverse of a valuation. In this case the analysis is to assess the single rate PV £1 pa used in the valuation which resulted in the sale price OR the analysis is to determine the investment's internal rate of return.

The internal rate of return using a financial calculator is 8.002%. For valuation purposes, this could be taken as 8%.

Proof

| | |
|---|---|
| Profit rent | £30,000 |
| PV £1 pa for 20 years at 8% | 9.8181 |
| Estimated market value | £294,500 |

### Example 6.2 A

A leasehold investment has just been sold for £325,000. The lease is of retail premises in a popular shopping street. The lease has 20 years to run without review, and the property has just been sub-let for the balance of the term, less one day, with rent reviews every five years. The current net profit rent is £30,000. Analyse to find the yield.
Using trial and error

| | |
|---|---|
| Profit rent | £30,000 |
| PV £1 pa for 20 years at x% | 10.83 |
| Sale price | £325,000 |

The yield which is closest to producing a multiplier of 10.83 is 6.75% (10.8). So, one can reasonably assume that this is the market yield for this type of leasehold investment.

Notes:

1. In Example 6.2, Goal Seek in Excel was used to find the internal rate of return at 8.002% and then 8% was substituted in a valuation to see if it produced a value approximating to the sale price; which of course it does.

2. In Example 6.2 A, the multiplier was found by dividing £325,000 by £30,000: this came to 10.83. Trial and error was then used by substituting in the PV £1 pa equation % until at 6.75% a figure of 10.80 was found. No further analysis was undertaken but this could have been refined to 6.751%, 6.752%, etc. - again Goal Seek would have produced the exact internal rate of return.

3. The reason different prices could occur in the market for leaseholds with the same current profit rent at the date of valuation and the same lease length is because one is a fixed profit rent and the other has the possibility of increases in the profit rent in five, 10 and 15 years time.

There is no way that the sale price of a leasehold investment can be analysed to determine a dual rate objectively, unless the analyst makes a prior assumption as to the accumulative rate. This is a first fundamental weakness in the dual rate approach. It cannot be based on objective market analysis; it can only be based on the valuer's own presumptions about capital recovery which are practiced by the buyers, hence the argument for analysing single rate and for valuing single rate.

If a single rate approach is to be used in a marketplace where buyers are liable for tax on incomes then capital recovery must be provided for out of taxed income. Investment advice can be provided on a before tax or after tax basis. In this example the net of tax return with tax at 40% is based on a net income (profit rent) of £18,000 (£30,000 minus tax at 40%). The net of tax return is only 1.99%. The reason for this low figure is because of the terminable nature of the investment and because there is no distinction for tax purposes between income which is in effect the return of the initial capital invested and the income which is the genuine investment income (see Chapter 2). To value single rate on a net of tax basis requires the valuer to either adjust the inherent sinking fund element or deduct tax from the profit rent. The latter approach is adopted in Example 6.3.

### Example 6.3

Estimate the present worth of a profit rent of £100 for four years. Tax is at 40% and the investor requires a net of tax return of 3.5%.

| | |
|---|---|
| Profit rent | £100 |
| Less tax at 40% | £40 |
| Net of tax profit rent | £60 |
| PV £1 pa for 4 years at 3.5% | 3.6731 |
| | £220.39 |

The investor will obtain a return on his or her capital of 3.5% net and will recover his or her capital in full at 3.5%. Whether or not part of the income is reinvested is immaterial, as the investor has the opportunity of accepting a partial return of capital at the end of each year.

| End year | Capital outstanding | Net income | Return on capital at 3.5% | Return of capital |
|---|---|---|---|---|
| 1 | £220.39 | £60 | £7.7136 | £52.29 |
| 2 | £168.10 | £60 | £5.88 | £54.12 |
| 3 | £113.98 | £60 | £3.99 | £56.01 |
| 4 | £57.97 | £60 | £2.03 | £57.97 |
| Total recovered | | | | £220.39 |

Most market valuations are undertaken on a before tax basis so it is essential to have due regard to the net of tax implications of investing in leaseholds when advising taxpaying investors. This is not a market valuation for the simple reason that it has been calculated at a specific tax rate; various rates of tax are payable from 0% to 50% and this is one reason why sale prices are analysed on a before tax basis and valuations are undertaken on a gross basis (see Chapter 7).

One of the critical issues about the use of dual rate was the realisation that many leasehold investments were being purchased by non-taxpayers, whilst the valuers who were advising sellers of leaseholds were valuing dual rate adjusted for tax. The full impact of this can be reviewed in Baum and Crosby (2008). The point here is that where any market is available to non-taxpayers, it is the taxpayer who suffers and has to accept the poorer net of tax return. Hence analysing the market and valuing on a before tax basis would seem to be appropriate for valuation purposes, but advising on a net of tax basis is essential for the taxpaying investor.

Single rate valuation of leaseholds poses no added issues beyond those already discussed in Chapter 5 on freehold valuations. It can comfortably deal with situations akin to the freehold term and reversion.

**Example 6.4**

Value the leasehold investment in a retail outlet in a popular shopping street. The lease has 20 years to run at a fixed head rent of £50,000. The current sub-lease has five years to run to the lease renewal at a rent of £80,000 and the market rent is £100,000. The appropriate leasehold capitalisation rate is 8% based on market comparables.

| Term | | |
|---|---|---|
| Rent receivable | £80,000 | |
| Rent payable | £50,000 | |
| Profit rent | £30,000 | |
| PV £1 pa for 5 years at 8% | 3.9927 | £119,781 |
| Reversion | | |
| Rent receivable | £100,000 | |
| Rent payable | £50,000 | |
| Profit rent | £50,000 | |
| PV £1 pa for 15 years at 8% | 8.5595 | |
| | £427,975 | |
| PV £ in 5 years at 8% | 0.68058 | £291,271 |
| Market value | | £411,052 |

The main issue facing valuers who are undertaking genuine lease-hold investment valuations is the lack of market comparable evidence. Every leasehold interest in property is unique in terms of lease, so where market analysis is available, valuers will frequently find themselves having to make adjustments to overcome differences between the comparable and the subject property. The other issues are the same and require a thorough inspection of the property and documents and clearly the valuation can be based on an in arrears assumption or quarterly in advance. The valuer may even need to deal with situations where the head rent is quarterly in advance and the sub-lease rents are monthly in advance. Careful application of basic principles will resolve all these problems.

## DCF valuation of leaseholds

DCF appears to be preferred by investors in leasehold properties, as only through DCF analysis can the investor begin to discover the possible benefits of buying a specific leasehold interest. DCF should therefore be the valuers' preferred method of assessing the market value or asking price or, for auction purposes, the reserve price. The reason why DCF may be preferred can be seen from the following example.

### Example 6.5

Value the leasehold interests in two retail units. A is a lease for 20 years at £10,000 pa without review and sub-let with five-year reviews at £20,000 pa; and B which is a similar lease for 20 years but of a larger unit held at a rent of £80,000 pa and sub-let at £90,000 pa with five-year reviews. Both are producing profit rents of £10,000 pa. Market evidence suggests leasehold capitalisation rates of 10%.

### Conventional single rate solution

| A | | B | |
|---|---|---|---|
| Profit rent | £10,000 | Profit rent | £10,000 |
| PV £1 pa for 20 years at 10% | 8.5136 | PV £1 pa for 20 years at 10% | 8.5136 |
| | £85,136 | | £85,136 |

Both leaseholds have been valued at £85,136, but what happens if occupation rents increase by 3% a year between the date of valuation and the first review in five years time? The rent will now be for A £23,185 (£20,000 plus 3% for five years compound) and for B the rent is £104,334 (£90,000 plus 3% for five years compound) and the values with 15 years to run at 10% become:

| A | | B | |
|---|---|---|---|
| Profit rent | £13,185 | Profit rent | £24,334 |
| PV £1 pa for | 7.6061 | PV £1 pa for | 7.6061 |
| 15 years at 10% | | 15 years at 10% | |
| | £100,286 | | £185,087 |

A has increased in value by £15,150, or by 17.8% and B has increased in value by £99,951, or by 117.4%. It is obvious on these assumptions that B is the much better investment and competition between investors might well push the price today up above the £85,136. However, this would be on the assumption that investors and their property advisors could see and would be aware of the potential in B if occupation rents increased. Meanwhile, there is a seller who may be being advised that £85,000 to £100,000 would be a good price for B.

Is there a better solution? The answer may be DCF. The two leases are comparable in risk but not in terms of potential for absolute growth. Moving the growth into the income flow and discounting at an appropriate target rate might resolve some of these points.

## DCF solution

The occupation rents will be increased by the market informed growth rate of 3% pa. The target rate will be built up logically from a gilt rate of 4.5%.

| | |
|---|---|
| Gilt Rate | 4.5% |
| Property premium | 2.5% |
| Leasehold risk | 2.0% |
| Sector risk | 0.5% |
| Location risk | 0.5% |
| Building risk | 0.5% |
| Tenant risk | 0.5% |
| Target | 11.0% |

| End of Year | Market rent plus 3% | Rent payable | Profit rent | PV £1 at 11.0% | Present value |
|---|---|---|---|---|---|
| 1 | £20,000 | £10,000 | £10,000 | .9009 | £9,009 |
| 2 | £20,600 | £10,000 | £10,000 | .8116 | £8,116 |
| 3 | £21,218 | £10,000 | £10,000 | .7312 | £7,312 |
| 4 | £21,855 | £10,000 | £10,000 | .6587 | £6,587 |

| End of Year | Market rent plus 3% | Rent payable | Profit rent | PV £1 at 11.0% | Present value |
|---|---|---|---|---|---|
| 5 | £22,510 | £10,000 | £10,000 | .5935 | £5,935 |
| 6 | £23,185 | £10,000 | £13,185 | .5346 | £7,049 |
| 7 | £23,881 | £10,000 | £13,185 | .4817 | £6,351 |
| 8 | £24,597 | £10,000 | £13,185 | .4339 | £5,721 |
| 9 | £25,335 | £10,000 | £13,185 | .3909 | £5,154 |
| 10 | £26,095 | £10,000 | £13,185 | .3522 | £4,644 |
| 11 | £26,878 | £10,000 | £16,878 | .3173 | £5,355 |
| 12 | £27,685 | £10,000 | £16,878 | .2858 | £4,824 |
| 13 | £28,515 | £10,000 | £16,878 | .2575 | £4,346 |
| 14 | £29,371 | £10,000 | £16,878 | .2320 | £3,916 |
| 15 | £30,252 | £10,000 | £16,878 | .2090 | £3,527 |
| 16 | £31,160 | £10,000 | £21,160 | .1883 | £3,984 |
| 17 | £32,095 | £10,000 | £21,160 | .1696 | £3,589 |
| 18 | £33,057 | £10,000 | £21,160 | .1528 | £3,233 |
| 19 | £34,049 | £10,000 | £21,160 | .1377 | £2,914 |
| 20 | £35,071 | £10,000 | £21,160 | .1240 | £2,624 |
| Present value | | | | | £104,190 |

If the assumptions here of 3% growth and 11.0% discount are correct then the capitalisation rate of 10.0% was incorrect. If the capitalisation rate of 10.0% was correct then the assumptions here are incorrect. The point is that the valuer is no longer hiding behind a mystery, but is now having to support the assumptions made – why 3% growth? Why 11.0% discount rate? Once more the reader can see how useful a spreadsheet would be for such an exercise and how important it is to think critically about the future.

For property B, the DCF table has been shortened because the task is broken down into four tranches of income and the PV £1 pa can be used as there are four constant cash flows at five-year intervals.

| Years | Market rent + 3%pa | Rent payable | Profit rent | PV £1 pa for 5 years at 11% | PV £1 | Present value |
|---|---|---|---|---|---|---|
| 1-5 | £90,000 | £80,000 | £10,000 | 3.6959 | 1 | £36,959 |
| 6-10 | £104,334 | £80,000 | £24,334 | 3.6959 | .5935 | £53,377 |
| 11-15 | £120,953 | £80,000 | £40,953 | 3.6959 | .3522 | £53,308 |
| 16-20 | £140,218 | £80,000 | £60,218 | 3.6959 | .2090 | £46,514 |
| Present value | | | | | | £190,158 |

The question is which method is the right method? Which values are the right values? Is the value of B really double the figure first suggested? Investors are likely to use a DCF, and to run scenarios and sensitivity analysis to establish an investment value that is the worth to them of a given leasehold investment. Concepts of worth are explored further in Chapter 12. However, it is the inefficiencies of the property market that make it so attractive to certain investor groups. If valuers are valuing having regard solely to today's profit rent then investments may be being undervalued. This would mean that leaseholds at certain times, as they have been, may be very attractive investments for some buyers. What valuers need to do is to pose the 'what if' question; to consider the probability of each outcome and then to review their opinions of value if DCF suggests that prices in the market might go above a capitalised profit rent if certain possible outcomes occur.

Earlier, an example of a leasehold investment was described where the head rent was fixed at 27.5% of the market rent. In other cases the head rents are reviewed every five years, in line with the occupation leases, but on a ground rent basis rather than a geared basis. In other situations the valuer may find that the head rent is reviewed every 25 years and the occupation leases every five years. Here it is suggested that a DCF approach would appear to be a logical way of setting out and discounting the probable profit rents. Profit rent x PV £1 pa is unlikely to expose the income characteristics of the more complex leasehold investments. This market is complex and the valuer will need to be as objective as possible.

## Summary

- Leases have value if there is a profit rent.
- The profit rent is the difference between the rent paid and the rent received or, in some cases, between market rent and rent paid.
- The profit rent must be assessed having regard to lease terms so that a net figure is derived, i.e. after outgoings; adjustments may be necessary if the head lease is internal rate of return and the property has been sub-let of internal terms only.
- The profit rent can be capitalised for the appropriate period of time, say for the unexpired term of the lease or to the point in time when head rent and rent received are the same, at a market capitalisation rate.
- Sales of leasehold investments can only be analysed to find their internal rate of return, any other analysis is subjective not objective.
- The leasehold valuation process is complicated by the fact that each lease is unique and may display a unique income pattern; a DCF approach may have to be adopted to reflect the unique pattern.

- Occupation lease pricing for assignment must be distinguished from investment lease valuation.
- Some valuers value leasehold investments on a dual rate basis (see Appendix A); this appears to be UK centric and the arguments for such an approach are not well founded; dual rate is not recommended by this text.
- Tax for those liable to tax will seriously impact on the return from leasehold investments due to the effect it has on that part of the profit rent which, like a loan, represents capital recovery to the investor; net of tax analysis of investment return should precede any decision to buy a leasehold.

# Chapter 7

## The Income Approach: Taxation and Valuation

## Introduction

It is customary in valuation to ignore the effects of income tax, on the grounds that investors compare investments on the basis of their gross rates of return. This may well be an acceptable criterion where tax affects all investments and all investors in a like manner. Although this is not the case in the property market, only a few valuers would argue that tax should always be deducted from income before being capitalised at a net-of-tax capitalisation rate (a *'true net approach'*).

One of the most important points to note is that where the income from property is all return on capital (that is, true spendable income), gross and net valuations will produce the same value estimate. Where part of the income is a part return of capital this, other than in the case of certain life annuities, is not exempt from taxation and in such cases the gross and net approaches may produce a different value estimate.

In addition, certain investments will produce fairly substantial growth in capital value over a relatively short term, due to a growth in income. In these cases, if the investment is resold, Capital Gains Tax (CGT) may be payable on the gain realised.

---

### Example 7.1

Explain what is meant by *'net rates of interest'* and discuss their uses in valuation using numerical examples.

A net rate of interest is interest earned on deposited monies after the deduction of an allowance for the payment of tax on the gross interest earned. Here the phrase refers to any net-of-tax rates.

Thus, £1,000 deposited with a bank, for example, earning interest at the rate of 10% per annum would, with tax at 40p in the £ (40%), produce a net rate of 6% (10 x (1 − t)) where t is 0.4. As far as the valuer's use of the financial mathematics is concerned, the switch from gross to net valuations should cause no problems. Thus, the present value of £1 in five years at 10% gross allowing for tax at 40p can be calculated by making the tax adjustment to the yield and using the net of tax rate in the formula. The PV of £1 at 10% adjusted for tax at 40% is the PV £1 at 6%.

It is also obvious that, where an investor is a taxpayer paying tax on all investment income at 40% or at the top rate of 50%, the actual return after tax will be reduced. The significance of this is not lost on investors, who always have full regard to their after tax returns. However, the question here is whether valuations should be undertaken on a before tax or after tax basis.

For simplicity of illustration a tax rate of 50% is used in some of the examples.

## Incomes in perpetuity

The formula for capitalising an income in perpetuity is income divided by $i$.

If the gross income is £1,000 and the gross capitalisation rate is 10% then: £1,000 ÷ 0.10 = £10,000

If tax is payable at 50p in the £ then:

$$£1,000 \times \left(\frac{1-t}{1}\right) \div 0.10 \times \left(\frac{1-t}{1}\right) = 500 \div 0.05 = £10,000$$

Clearly, as the numerator and denominator are multiplied by a constant $\left(\frac{1-t}{1}\right)$ then they can be divided through by that constant, hence:

$$I \div i = \left(\frac{I \times \left(\frac{1-t}{1}\right)}{i \times \left(\frac{1-t}{1}\right)}\right)$$

No difference in value estimate will occur, because the income is perpetual and is all return on capital.

## Finite or terminable incomes

An income receivable for a fixed term of years may have a present value which can be assessed on a gross or net of tax basis.

Consider an income of £1,000 receivable for five years on a 10% gross basis with tax at 50%.

|  | Gross | | Net of tax at 50p |
|---|---|---|---|
|  | £1,000 | | £500 |
| PV £1 pa for 5 years at 10% | 3.7908 | PV £1 pa for 5 years at 5% | 4.32955 |
|  | £3,790 | | £2,164 |

Here, there is a clear difference between the two figures which can be seen to result from the tax adjustments that have been made:

$$I \times \left(\frac{1 - \dfrac{1}{(1+i)^n}}{i}\right) \text{ cannot be equated with } I(1-t) \times \left[\frac{1 - \dfrac{1}{(1+i(1-t))^n}}{i(1-t)}\right]$$

Here, part of the £1,000 income is a return of capital. If tax is payable at 50p in the £, then only £500 is available to provide a return on capital and a return of capital. A purchase at £3,790 would be too high to allow for the returns implied if tax is at 50p. Where tax is payable, replacement of capital (purchase price) must be made out of taxed income.

In order to preserve the gross rate of 10% for investment comparison the valuation can be reworked, recognising that the return on capital is at 10% and the return of capital is at a net rate out of taxed income. $i$ is therefore 0.10 and the annual sinking fund (ASF) in the formula is at 5% adjusted for tax.

$$£1,000 \times \frac{1}{0.10 + \left(0.1809 \times \dfrac{1}{1-t}\right)} = £2,165$$

Or, set out as a valuation:

|  |  |
|---|---|
|  | £1,000 |
| YP for five years at 10% and 5% adj. tax at 50%* | 2.165 |
|  | £2,165 |

$$^{*}\text{Formula} = \frac{1}{i + ASF} = \frac{1}{0.10 + \left(0.1809\left(\dfrac{1}{1-t}\right)\right)} = \frac{1}{0.10 + 0.3618} = 2.165$$

The gross valuation has been adjusted to equate with the net valuation achieving a 10% return on capital and a return of capital out of taxed income.

| Capital outstanding | | 10% gross | Capital recovered (£1,000 - 10% - 50p in £) tax |
|---|---|---|---|
| 1 | 2,165.00 | 216.500 | 391.75 |
| 2 | 1,773.25 | 177.325 | 411.3375 |
| 3 | 1,361.91 | 136.190 | 431.90 |
| 4 | 930.01 | 93.000 | 453.50 |
| 5 | 476.50 | 47.650 | 476.175* |
|  |  |  | £2,165 |

*Error due to rounding.

However, whilst this sets out how to equate a gross valuation (10%) with a net valuation (5%), it does not answer the questions 'should property be valued on a before tax basis or after tax basis?' 'Where does the 10% come from and should the value be £3,790 or £2,165?'

Property provides an investment opportunity for all investors. Individuals with sufficient capital to buy property are likely to be liable to tax on income at the highest rate which is now 50%. Some organisations, such as charities, Real Estate Investment Trusts (REITs) and Pension Funds may be exempt from income tax, whilst certain corporate investment companies will be taxed. Valuing on any basis other than by analysing sales on a before tax basis and valuing on a gross basis would be valuing to a class of investor and as such would not be a market valuation but an assessment of investment value or worth to that class of investor.

## Tax and deferred incomes

A deferred income is one due to commence at a given date in the future.

### Example 7.2

Calculate the present worth of an income of £2,000 per annum in perpetuity due to commence in five years' time. A 10% gross return is required, and tax is payable at 50%.

|  | Gross |  | Net |
|---|---|---|---|
|  | £2,000 |  | £1,000 |
| PV £1 pa perp. at 10% | 10 | PV £1 pa perp. at 5% | 20 |
|  | £20,000 |  | £20,000 |
| PV £1 in 5 years at 10% | 0.6209 | PV £1 in 5 years at 5% | 0.7835 |
|  | £12,418 |  | £15,670 |

Here it can be seen that capitalisation in perpetuity gross and net produce the same value of £20,000 in five years time, but, by further discounting to allow for the five years deferment, the gross and net valuations of the deferred income produce different results arising from the deferment factors of 0.6209 and 0.7835.

Obviously, a buyer would wish to pay only £12,418, but what of the seller?

Assuming initially that capital gains are not taxable, one can see that an investment purchased for £12,418 held for five years and sold for £20,000 is a 10% investment: £12,418 x amount £1 for 5 years at 10% = £20,000.

However, if an investor were to deposit £12,418 in an income producing investment, the 10% return of £1,241.80 in year one would be taxed. From a taxpayer's point of view, a 10% return, all in capital growth with no tax would be better than 10% all in income subject to taxation.

If there was no CGT, capital growth investments would be very attractive to high rate taxpayers, to such an extent that prices could be pushed up to the point of indifference, i.e. until the capital growth investment is equated with an income investment. In this example, if tax is at 50p in the £, a taxpayer would get the same return after tax from depositing £15,670 at 10% gross as from purchasing £2,000 pa in perpetuity deferred five years for £15,670.

The introduction of CGT in 1965 reduced the tax-free element of capital growth but did not lessen the belief held by some valuers that the net approach was still correct. It is recognised that the incidence of all taxes should be reflected in an investor's true return. Therefore, if tax is material to investors' decisions in the marketplace it should be allowed for in an investment valuation but not necessarily in an estimate of market value.

To explain the need for analysis net of tax and for net of tax investment valuations, a single question needs to be posed: 'If investors are assured of a 10% gross return and pay tax at 50%, are they expecting a 5% net-of-tax return?' To this, one could add a supplementary question: 'Are investors interested in their return net of tax?'

It must be assumed that, as HM Revenue & Customs has a prior claim on investment income, investors must have some regard to their net of tax income. However, this does not infer that valuations must be undertaken on a net of tax basis. Current market practice appears to be to value gross but, if requested, to advise clients, in consultation with their tax advisors, on a net of tax basis.

# Rising freehold incomes

Combining the £1,000 income for five years with the £2,000 income commencing in five years and continuing in perpetuity, the result will be a normal term and reversion valuation.

Using the figures from the preceding examples, the gross and net valuations produce these results:

|  | Gross | Net |
| --- | --- | --- |
| Value of £1,000 for 5 years | £3,790 | £2,164 |
| Value today of reversion to £20,000 | £12,418 | £15,670 |
|  | £16,208 | £17,834 |

(In such cases, the net valuation will always be higher than its gross equivalent.)

The gross valuation of £16,208 may be analysed in two parts: an expenditure of £3,790 to acquire £1,000 for five years, coupled with an expenditure of £12,418 for £20,000 in five years time. In the absence of income tax, the investor readily achieves his or her 10% return: but what of the investor's 5% net return?

After tax on income, the investment will have the following cash flow:

| Year | Cash flow |
| --- | --- |
| 1 | £500 |
| 2 | £500 |
| 3 | £500 |
| 4 | £500 |
| 5 | £20,500 |

This is the cash flow assumed for the 5% net valuation, and the investor would achieve a 5% return at an acquisition cost of £17,834. Therefore at £16,208, the investment must be showing a better return than 5%. If investors only expect a net return of 5%, valuing gross may result in vendors' interests being undersold.

## Gross funds

The term 'gross funds' is used today to describe any investor exempt from income, corporation and/or CGT. These are principally charities, local authorities, approved superannuation funds, pension funds, REITs, friendly societies and registered trade unions. Other institutions, such as insurance companies, enjoy partial relief by being assessed at a lower rate. Many first-class property investments are now held by gross funds, which are an increasingly large part of the property market – so much so that co-ownership schemes in agricultural investment and in commercial investment have been carefully created to allow small and large funds to buy, as co-owners, substantial single investments in property while still preserving their tax-exempt status.

If the most probable purchaser in the market is a gross fund then the sub-market is one of gross funds. As they pay neither income nor CGT, the return to them will be both their gross and net return. Valuations within this sub-market must be based on analysis of comparable transactions, which suggests that no adjustments should be made for tax.

## Net or gross?

As a difference in the value may result when valuing net or gross of tax, the problem therefore remains.

Should the valuation be net or gross?

There is no categorical answer: a number of points may, however, be made.

1. In most cases a net valuation will not produce an accurate estimate of market value, as every potential buyer may have a different tax liability.
2. A net valuation should be carried out when advising a buyer on their maximum bid for an investment when their required net of tax return is known.
3. In certain sub-markets the market is dominated by non-taxpayers. In these cases a gross valuation will produce the same valuation as a net valuation allowing for tax at 0%.

## CGT

The introduction of CGT in 1965 had an immediate impact on the property market. Initially it reduced the obvious benefit of investing in properties with high capital growth expectations, such as freeholds with early substantial reversions. The market soon adapted to the changes but, although there were pleas from a few to move to a net of tax approach to valuation, the market responded with a simple adjustment to the all risks yield (ARY) to account for this new risk element.

Reversionary properties remain more attractive to individual investors because of the differences between the CGT tax rate (28%) and the higher personal tax rates (40% and 50%). In the market for prime multi-million pound properties, the dominant buyers are the tax exempt funds; individuals wishing to invest here must outbid these funds and accept a low net of tax return. Further tax advantages for some investors of reversionary properties are as follows:

1. CGT is charged on net gains, i.e. after deduction of purchase and sale costs and allowable expenses.
2. CGT is charged on the gain after deducting the taxpayer's tax free allowance.
3. The tax demand may be up to two years after the gain is realised.
5. CGT may, in some cases, be deferred through roll over relief.

For these reasons, individuals with a 50% income tax liability may still favour the highly reversionary properties in preference to rack rented properties, as a substantial part of the total return will be the capital gain which will be taxed at 28% not 50%. However, it is probably the ability of the right property to outpace inflation that attracts the private investor.

For the major investors, because property is normally purchased as a long-term investment, any gain is likely to be deferred for many years and will not therefore have a significant bearing on price. In addition, 'rollover-relief' may be possible, effectively postponing any CGT and, in any event, for certain classes of investor their exemption from CGT makes them a market in their own right.

There may, however, be cases where property is to be bought with the intention of realising a capital gain at a specific date in the future. Where the buyers in the market are likely to have the same intent and the client requires advice on gross and net of tax returns, it might be necessary to reflect CGT in the calculations. The client's tax consultant must be involved in any net of tax advice offered.

# Value Added Tax

Many aspects of developing, owning, running and maintaining property including property transactions, other than exempt categories of property, incur Value Added Tax (VAT). This functions as a significant tax in those cases where it has to be paid and the payer is in a non-recoverable position. The assessment of market value is generally undertaken without specific regard to VAT. Where VAT may need to be considered is in the area of property investment advice where buying or letting on the wrong basis regarding VAT could be very costly. Details of this subject are outside the scope of this book.

## Summary

- Income capitalisation of property is usually undertaken on a before tax basis.
- Tax on income has no effect where all income is return on capital.
- Tax may need to be reflected where investments are terminable in the short term and tax is a factor in the market place.
- Tax is potentially significant when valuing leasehold investments in a market dominated by taxpayers.

# Chapter 8

## Landlord and Tenant

## Valuations and negotiations

The income approach to property valuation centres mainly on the estimation of market value, but the use of the same tools in resolving value problems between landlords and tenants is of particular interest. This chapter examines several of these value issues.

Negotiators need to understand the negotiating position of the other party and for this reason many of these issues require consideration from both the landlord's and the tenant's perspectives. Previous editions of this book and a number of other texts view this as equating value between current market value and future market value, i.e. post the negotiations. In considering both parties' positions, both interests have traditionally been valued 'before and after' in the context of the normal definition of market value. However, whilst freeholders will be concerned to ensure that the value of their freehold interest is not diminished as a result of any lease negotiations with an occupying or prospective tenant, tenants are occupiers or prospective occupiers of property, and their concern is to ensure that their occupation costs are not increased as a result of lease negotiations; for this reason strict adherence to concepts of market value 'before and after' for any tenant calculations might not be appropriate – the last thing a tenant would be contemplating in these circumstances would be marketing their occupation rights.

In addition, the negotiator needs to be aware of each party's negotiating strength in general and in particular. For example, revisions for this edition were at a time when, due to the economic situation, tenants could negotiate from a position of strength; but an individual tenant might nevertheless, due to their own circumstances, have to negotiate from a position of weakness.

## Premiums

A premium is a lump sum paid by a tenant to a landlord in consideration for a lease granted at a low rent, or for some other benefit. 'At a low rent' signifies a rent below market rent, and the other benefits will be as a rule, financial, having the same effect as a reduction in rent.

Examples of this are the tenant paying for immediate repairs that would normally be the landlord's responsibility, or financing an improvement without being charged an increase in rent.

A premium is often paid on the grant or renewal of a lease, but there may be more than one premium payable at any time during the lease term. It entails a cash gain, coupled with a loss of rent for the landlord because the usual result of charging a premium will be a letting at less than market rent; effectively selling part of the freehold. The tenant will be paying a cash sum in return for a lease at a rent below market rent, effectively buying part of the freehold and securing a profit rent.

This can be illustrated by looking at the relationship between market rent and market value. Given a market rent of £10,000 and a market capitalisation rate of 10%, the market value of a freehold interest in a property would be £100,000. A sale represents the full disposal by the owner of all rights to any part of the market rent. However, if the owner only wished to sell part of his or her entitlement, the owner could effectively sell five, 10, 15 or however many years of the freehold title to all of the property, or they could sell part of his or her right to the £10,000 a year rent. A premium is, in effect, a part disposal of the freehold to a tenant and may give rise to a charge to Capital Gains Tax (CGT).

The following table illustrates the premium an owner would require in lieu of an increasing reduction of market rent over a five-year term.

| Reduction in rent | Years Purchase for 5 years at 10% | Premium required at 10% to nearest £1 |
| --- | --- | --- |
| £1,000 | 3.7908 | £3,791 |
| £2,000 | 3.7908 | £7,582 |
| £3,000 | 3.7908 | £11,372 |
| £4,000 | 3.7908 | £15,163 |
| £5,000 | 3.7908 | £18,954 |

The explanation is simple in that for every £1,000 in rent reduction over the five-year term, the market value of the owner's freehold interest falls immediately by £3,790.80.

### Why pay a premium?

A premium has some advantages to a landlord but few for a tenant. Premiums are often paid in a seller's market, i.e. where there is competition among prospective tenants to secure an agreement with the prospective landlord.

The advantages to a landlord are several:

1. Although the amount of the premium will reflect the discounting of future income, an immediate lump sum receivable instead of a future flow of income is often more attractive due to the 'time value of money'. Additionally, the landlord may prefer a lump sum in order to meet an immediate expense or to make any kind of cash investment.

2. Receipt of a lump sum immediately may reduce the diminishing effect that inflation has on the value of future income in real terms, especially if rent review periods are longer than is favourable to the landlord.

3. A premium may have tax advantages, if the landlord is a taxpayer, due to the different treatment of income and capital gains for taxation purposes.

4. A premium should increase the landlord's security of income. Once the tenant has paid a premium and invested money, the tenant is more likely to remain in occupation of the premises and should be a more reliable tenant. Some of the risk of non-payment of rent may be reduced.

The advantages to a tenant are less well-defined:

1. A premium may be useful as a loss or deduction to be made from profits when being assessed for income tax or CGT.

2. Paying a premium may be advantageous to a tenant when it is preferable to the tenant to part with capital in order to reduce future recurring expenses.

However, the landlord will usually enjoy the greater benefits.

When a property attracts many prospective tenants, a landlord may demand and receive a premium in addition to rent. Valuers must be careful to note that this represents the capital value of the extra rent, above market rent, that such excess competition can sometimes generate. The following section deals with the assessment of premiums for rents forgone.

## Valuation technique

A premium entails a loss to the landlord of part of the income and a gain to the tenant of a profit rent. The amount of the premium should be calculated so that each party is in virtually the same position as if market rent were to be paid and received.

The gain or loss of rent capitalised over the period for which it is applicable should be calculated to be equal to the amount of the premium. It is conventional practice to use full freehold rates to capitalise the landlord's loss of rent and full leasehold rates to capitalise the tenant's gain of profit rent, presumably on the basis that each party by definition is to be in the same position as if market rent were being paid and received. Such a conventional approach from a leaseholder's point of view could be criticised where the valuer uses a dual rate or dual rate adjusted for tax approach. The whole argument put forward for using dual rate is based on a false premise about capital recovery. It needs to be understood that where a premium is being paid in lieu of rent, it is simply a time value money exchange and the calculations should only be undertaken on a single rate basis. In many situations this only really needs to be undertaken from a landlord's point of view because leases are deemed to be riskier investments and will be valued at a higher capitalisation rate, suggesting that a tenant would only offer a sum lower than the landlord's requirement. This would create a negotiation impasse. It would be more appropriate for the tenant to consider, with his or her accountant, the affect on pre- and post-tax profits of paying a premium rather than paying the full annual rent.

---

### Example 8.1

What premium should A charge on the grant of a 10-year lease without a rent review to B at a rent of £15,000? The market rent is £25,000. Assume for illustration a freehold rate of 10%.

*Therefore:*

| | | |
|---|---|---|
| A's loss of rent | £10,000 | |
| PV £1 pa for 10 years at 10% | 6.1446 | £61,446 |

A truer picture of real gain and real loss requires an explicit discounted cash flow (DCF) approach.

---

### Example 8.2

Demonstrate that the premium agreement in Example 8.1, from a 'before and after' perspective, provides the freeholder with the same value equivalence.

| MV* before | | MV after | | |
|---|---|---|---|---|
| MR** | £25,000 | Rent | £15,000 | |
| PV† £1 pa in perp at 10% | 10.00 | PV £1 pa for 10years at 10% | 6.1446 | |
| Value | £250,000 | | | £92,169 |
| | | | | |
| | | Reversion to MR | £25,000 | |
| | | PV £1 pa in perp defd10 years at 10% | 3.8554 | £96,385 |
| | | | | £188,554 |
| | | Plus premium | | £61,446 |
| | | | | £250,000 |

*Market value.
**Market rent.
†Present value.

This then is the value to the freeholder before and after but the deal being negotiated might impact on the market value after. So, for example, an investment might be currently seen as a 7% opportunity but the deal being negotiated with a tenant might create an investment opportunity, post the negotiations, which the market would regard as an 8% risk; in which case a 'before and after' set of calculations would require a higher premium than might be suggested by a simple capitalisation of the rent foregone. For this reason many valuation surveyors prefer to consider the 'before and after' position of the freehold market values rather than just the capitalised rent foregone.

Emerging in the 1990s and re-emerging in 2007 has been the concept of reverse premiums. Typically, an owner of property who is reluctant to appear to be letting the whole or part of a property below historic market rents would offer a premium or cash incentive to a tenant on signing a lease. Similarly, tenants occupying over-rented property have had to offer premiums when sub-letting the whole or part. The key to the valuation problem is to recognise that a reverse premium is a payment from a landlord to a tenant. The principles behind their calculation are similar and a 'before and after' valuation is recommended to reflect the market attitude to properties when the rent is being held artificially above normal market rents.

> **Question 1:**
>
> Calculate the premium to be paid when a shop property (5% ARY) is to be let at £50,000 on 5-year normal lease terms when the market rent on similar terms is £75,000.

## Future costs and receipts

A study of conventional techniques of deferring future costs raises a number of further questions which are examined shortly.

The deferment of future receipts is relatively straightforward. Often an investment will provide a capital sum at some given time in the future. An example of this is a premium payable during the currency of a lease. This is part of the investment and could therefore be discounted at the remunerative rate.

**Example 8.3**

A lets number 10 High Street to B for 21 years at £139,250 pa. In addition, a premium of £150,000 is payable at the end of the lease. MR is £150,000 pa. Value A's interest.

*Assuming a freehold rate of 10%*

| | | |
|---|---|---|
| Rent | £139,250 | |
| PV £1 pa for 21 years at 10% | 8.6487 | £1,204,331.48 |
| | | |
| Rent | £150,000 | |
| PV £1 pa in perp defd. 21 years at 10% | 1.35131 | £202,696.50 |
| | | |
| Premium | £150,000 | |
| PV £1 in 21 years at 10% | 0.13513 | £20,269.50 |
| Estimated capital value | | £1,427,297.48 |

This approach implies that future fixed receipts should be discounted at the market yield or remunerative rate and added to the freehold value; in which case the premium of £150,000 varies in value according to the quality of the investment. This conventional approach is supportable, in that the risk attaching to the receipt of the future premium is effectively the same as the risk attaching to receiving the rent, much depends on the tenant's covenant. However, in present value terms it could be argued that such future sums should not vary with the property but with money market rates. This 10%

equivalent yield approach might be difficult to justify from market evidence and, without resorting to a growth explicit DCF, some valuers would capitalise the fixed 21-year term at a high opportunity cost rate reverting to a market based capitalisation rate after 21 years and discounting the premium at the same opportunity cost rate.

Future capital costs will often arise out of investment in property. Conventional valuation practice distinguishes two types of future capital cost.

## Liabilities

These may have to be incurred for some reason and are a cost to the investor, often being legally enforceable. They must be allowed for in a valuation. Examples are sums for new plant, lifts, etc.

The investor must make certain that the cash required for the liability is available at the relevant time. For this reason typical savings rates should be used rather than the market yield appropriate to that property.

## Expenditure

This is an optional spend and will only be undertaken if it provides a sufficient return on the sum to be invested. This 'sufficient return' is taken for convenience to be the rate of return that the investment as a whole provides, and so expenditures are discounted at the market yield for that property.

The distinction between liabilities and expenditures may be valuable to investment decision taking.

---

### Example 8.4

B offers £15,000 to A for the leasehold interest. A receives a profit rent of £3,800 pa with 14 years remaining and no rent reviews on head- or sub-lease. A has agreed to carry out repairs at a cost of £10,000 in four years time. Should B's offer be accepted?

Assuming a leasehold rate of 14%:

| | | | |
|---|---|---|---|
| Profit rent | | £3,800 pa | |
| PV £1 pa for 14 years at 14% | | 6.0021 | £22,807.98 |
| | | | |
| Less: Cost of repairs | | | |
| (I) *Liability?* | £10,000 | | |
| PV £1 in 4 years at 2% | 0.9238454 | | £9,238.45 |
| Estimated capital value (I) | | | £13,569.53 say £13,570 |
| | | | |
| or | | | |
| (2) *Expenditure?* | £10,000 | | |
| PV £1 in 4 years at 14% | 0.592080 | | £5,920.80 |
| Estimated capital value (2) | | | £16,887.18 say £16,887 |

The distinction is vital; if the cost is treated as a liability then the offer is accepted but not if treated as expenditure. In many cases the decision is an arbitrary one, the effect of which can be reduced by the use of realistic 'safe' money rates.

A further problem now emerges, and that is the estimation of future expenditures and liabilities when they are not known, or fixed in monetary terms at the date of valuation. For example, an expected major renewal in four years time can only initially be estimated on the basis of cost today. However, the deduction to be made from market value in good condition must be an amount the market considers to fairly reflect the current condition. Where the future sum is fixed in monetary terms then a present value calculation at a realistic monetary safe rate is satisfactory. Where it is a current estimate then the valuer must consider whether the costs will increase at a faster rate than that which can be safely earned on savings.

If the expenditure is likely to rise at 3% a year and money can be saved to earn interest at 1% a year, then the wisest solution is to deduct the full cost from today's value. If money can earn interest at a higher rate than the estimated inflationary increase in costs, then a discounted sum can be deducted. If costs are rising faster than money rates, then it would be logical to deduct the full cost now, and indeed to have the repairs or renewals undertaken now. However, unless the work is essential as at the date of valuation, it may be as realistic to simply write down the value in sound condition by an appropriate amount.

## Extensions and renewals of leases

The occupier or tenant of business premises will often be anxious to remain in occupation because a business move might involve considerable expense, loss of trade and loss of goodwill, resulting in a large loss of business profit. In these circumstances tenants will be keen to negotiate an extension of the lease, a renewal of the existing lease on similar terms, or the grant of a completely new lease on different terms. Similarly, landlords anxious not to lose a tenant may commence negotiations for a new lease two to three years prior to the end of a lease and ahead of the statutory or contractual date for service of notice to terminate the lease. These situations will generally occur where the current lease was for a long term, and especially if for a long term without rent reviews. Leases today are getting shorter, typically 10 years, and in smaller towns often for no more than three years.

The problem of a tenant who approaches the landlord as a short lease draws to an end provides very little difficulty. The landlord will require a rent approaching or at market rent, and the tenant will expect to pay it because that is what it would cost to lease a comparable property. However, finding alternative accommodation is not an easy operation and business decisions are prudently taken well in advance. It is thus more usual for a tenant to approach the landlord well before the termination of the existing long lease and when they do, a valuation problem will arise. If the tenant approaches the landlord during the currency of the lease with a proposal to renew that lease immediately for an extended period, then it follows that the tenant is offering to surrender the current lease. (Such problems are often called 'surrender and renewals'.)

If, as is probably the case, a profit rent or a notional profit rent has arisen, then any surrender will be a surrender of valuable leasehold rights in the property. The tenant would be ill-advised to accept a new lease at market rent.

The landlord, on the other hand, is not likely to agree to any indiscriminate extension of the tenant's profit rent because the anticipated reversion to market rent will already be reflected in the market value of the freehold. Negotiations must be conducted to see that, following the surrender and lease renewal, there is no diminution in the value of the landlord's or of the tenant's interests in the property.

Valuers acting for the two parties, or acting as arbitrator or independent expert will be checking the position from both sides, on the basis that the value of the present interests should equal the value of the proposed interests. This will involve four or more valuations.

---

## Example 8.5

A tenant occupies premises on a 21-year lease with two years to run at £15,000 a year. The tenant wishes to surrender this lease for a new lease for 15 years, with rent reviews every five years. The MR on full repairing and insuring (FRI) terms for 15 years with five-year reviews is £30,000. Freehold all risks yield (ARY) is 7%.

| Present Interest of the tenant | | | Present Interest of the landlord | | |
|---|---|---|---|---|---|
| MR | £30,000 | | Current rent | £15,000 | |
| Rent payable | £15,000 | | PV £1 pa 2 years at 7% | 1.8080 | £27,120 |
| | | | | | |
| Profit rent | £15,000 | | | | |
| PV £1 pa 2 years at 10% | 1.7355 | | MR | £30,000 | |
| Value | £26,032 | | PV £1 pa perp defd 2 years at 7% | 12.4777 | £374,331 |
| | | | | | £401,451 |

| Proposed Interest of the tenant | | | Proposed Interest of the landlord | | |
|---|---|---|---|---|---|
| MR | £30,000 | | Rent | £x | |
| Rent payable | £x | | PV £1 pa 5 Years at 7% | 4.1002 | £4.1002x |
| | | | | | |
| Profit rent | £30,000−x | | MR | £30,000 | |
| PV£1 pa 5 years at 10% | 3.7908 | | PV £1 pa perp defd 5 years at 7% | 10.1855 | £305,565 |
| | £113,724 − 3.7908x | | | | £305,565+ £4.1002x |

Note that it is both the rent and the £x new rent that have to be capitalised.

Equating present with proposed interests:

**Tenant's position**

| £26,032 | = £113,724 − £3.7908x |
| £3.7908x | = £87,692 |
| £x | = £87,692/3.7908 |
| £x | = £23,132 |

**Freeholder's position**

£401,451 = £305,565 + £4.1002x

£401,451 - £305,565 = £4.1002x

£95,886 = £4.1002x

£x = £95,886/4.1002 = £23,385

The landlord requires a minimum rent of £23,385, while the tenant can afford to offer £23,132. Negotiation between the parties will take place. A settlement at around £23,200 would seem to be likely.

It must be pointed out that the above type of solution may result in a tenant's bid being lower than the landlord's minimum requirement. If the gap is large, the proposals may be shelved. Considerable forces of inertia will, however, normally conspire to produce an agreement if the shortfall is of a minor nature; for example, the landlord, if satisfied with the tenant, may wish to save the advertising and legal fees involved in finding a new tenant. In addition, the current tenant will have many reasons, as already discussed, for being prepared to make a small loss in order to carry on in occupation of the premises.

### Example 8.6

T occupies 6 High Street, holding a lease from L at a rent of £20,000 pa FRI with eight years remaining. T requires a new 25-year lease, starting immediately, and proposes to carry out improvements to the premises in three years' time at an estimated cost of £120,000 which will increase market rent by £25,000 pa. The market rent is £80,000 pa on FRI terms with rent reviews every five years. Comparable freeholds are selling at 9% and leaseholds at 11%. Here, it is important to consider whether or not it is in the landlord's best interests to meet the costs of the improvements at the end of the lease and recover the new improved market rent of £105,000, or let the tenant carry out the works and accept, in accordance with normal landlord and tenant relationships, that the licence to carry out the improvements or the conditions in the new lease will preclude the landlord from charging rent on the tenant's improvements for a minimum of 21 years from practical completion (see Chapter 9 for statutory regulation on this point under the Landlord and Tenant Act 1954 Part II as amended but note that here the negotiations are unfettered).

Acting between the parties, assess what rent should be fixed under the proposals.

### A Landlord's present interest

| Rent | £20,000 |
| PV £1 pa for 8 years at 9% | 5.5348 |
| | £110,696 |

Reversion (1):

Assuming the landlord does not carry out the proposed improvements.

| | |
|---|---|
| Rent | £80,000 |
| PV £1 pa perp defd 8 years at 9%* | 5.5763 |
| | £446,104 |

Value with no improvements is £110,696 plus £446,104 which is £556,800.

Reversion (2):

Assuming landlord does carry out the improvements in eight years time, and assuming the delay is short enough to involve no appreciable loss of rent.

| | |
|---|---|
| Rent | £105,000 |
| PV £1 pa perp defd 8 years at 9% | 5.5763 |
| | £585,511 |

*Both rents are market rents, although they differ in magnitude, the security of income is assumed unchanged.

The figure of £585,511 can only be achieved by an expenditure of £120,000 in eight years time and therefore value today must be reduced accordingly.

| | |
|---|---|
| Cost of improvements | £120,000 |
| PV £1 in 8 years at 9%* | 0.5019 |
| | £60,228 |

*This is 'expenditure' and discounted (possibly erroneously) at the market yield.

This leaves a net present value for the reversion of £525,283 (£585, 511 − £60, 228).

Reversion (2) is more valuable so the landlord would improve, and the value of the freehold interest is this increased sum.

Estimated capital value
£110,696 + £525,283 = £635,979

### B Landlord's proposed interest

| | | | |
|---|---|---|---|
| Rent | | £x | |
| PV £1 pa for 5 years at 9% | | 3.8897 | 3.8897x |
| | | | |
| Reversion to unimproved MR* | | £80,000 | |
| PV £1 pa for 20 years at 9% | 9.1285 | | |
| PV £1 in 5 years at 9% | 0.6499 | 5.9326 | £474,608 |
| | | | |
| Reversion to MR improved* | | £105,000 | |
| PV £1 pa perp defd 25 years at 9% | | 1.2885 | £135,292 |
| | | | £609,900 + £3.8897x |

*The landlord's proposed interest is being assessed in the light of the agreed terms for the new lease. In this example this means that the landlord will give the tenant 21 years enjoyment of the improvements paid for by the tenant. Effectively until the end of the 25 year lease. At the intermediate rent reviews there will be increases to the unimproved MR.

Present interest = proposed interest

$$£635,979 = £609,900 + £3.8897x$$
$$£635,979 - £609,900 = £3.8897x$$
$$£26,079 = £3.8897x$$
$$£6,705 = £x$$

This means that if agreement is to be reached, the landlord in recognition of the tenant's current profit rent will accept c £6,705 for the first five years, rent reviews to the unimproved MR at the fifth 10th and 15th year with a review to the improved MR after 25 years and thereafter. Technically this might occur after 18 years (21 years less the three years before the improvements are completed), but a lease arranged on this basis might be difficult to negotiate. Conventional valuation practice is to reflect any changes in the future rents by using a capitalisation rate which is market based and accounts for all risks be they negative or positive. The alternative would be a full DCF.

### C Tenant's present interest

| | | | |
|---|---|---|---|
| Rent received | £80,000 | FRI | |
| Rent paid | £20,000 | FRI | |
| Profit rent | £60,000 | | |
| PV £1 pa for 8 years at 11% | 5.1461 | | £308,766 |

### D Tenant's proposed interest

| | | |
|---|---|---|
| Rent received | £80,000 | |
| Rent paid | x | |
| Profit rent | £80,000 − £x | |
| PV £1 pa for 3* years at 11% | 2.4437 | £195,496 − £2.4437x |

*After 3 years T improves the premises and increases MR.

| | | | | |
|---|---|---|---|---|
| Reversion to: | | | | |
| Rent received | £105,000 | | | |
| Rent paid | x | | | |
| Profit rent | £105,000 − £x | | | |
| PV £1 pa for 2 years at 11% | 1.7125 | | | |
| PV £1 in 3 years at 11% | 0.7312 | 1.2522 | £131,479 − £1.2522x |

| | | | | |
|---|---|---|---|---|
| Reversion to: | | | | |
| Rent received | £105,000 | | | |
| Rent paid | £80,000 | | | |
| Profit rent | £25,000 | | | |
| PV £1 pa for 20 years at 11% | 7.9633 | | | |
| PV £1 in 5 years at 11% | 0.5935 | 4.7262 | £118,155 |

£445,130 − £3.6959x

Adding the constituent elements of the valuation together produces £445,130 − £3.6959x:

Less

| | | |
|---|---|---|
| Improvements | £120,000 | |
| PV £1 in 3 years at 11% | 0.7312 | £87,744 |
| New total value: | £445,130 − £3.6959x − £87,744 = £357,386 − £3.6959x | |

| Present interest | = Proposed interest |
|---|---|
| £308,766 | = £357,386 − £3.6959x |
| £3.6959x | = £357,386 − £308,766 |
| £3.6959x | = £48,620 |
| £x | =£13,155 |

| Minimum L will accept: | £6,705 pa |
|---|---|
| Maximum T can offer: | £13,155 pa |
| Agreement between parties of (say): | £10,000 pa |

As the tenant's maximum bid is above the landlord's minimum acceptable figure, an agreement is very likely. The reasons for the differences lie in the use of the different capitalisation rates of 9% and 11% and the earlier increase in profit rent enjoyed post improvement by the tenant. Critical in such considerations are the underlying assumptions of each calculation – are they realistic? The valuer is likely to be aware of market yields (cap rates) for freehold interests, but less likely to have direct evidence of comparable lease rates. Typically in these cases, a yield 2% to 4% above the equivalent freehold rate might be used. The results will differ and on occasions it will appear that the tenant's view is below the landlord's view and no settlement would be possible. The rational argument here as with the premium is that the landlord's position should prevail but there is no answer until the parties begin their negotiations.

The valuer must have regard to, and take account of, all the peculiarities of conventional valuation, the problems of future liabilities entered at current costs, the problems of long reversionary leases without rent reviews and the way in which a surrender and renewal can alter the market's perception of both the freehold and leasehold interests.

### Question 2: Surrender and renewal

An office building was let 30 years ago for 40 years without a rent review clause in the lease, at £10,000 a year. It is now estimated that the building would let in the market at £100,000 a year. Similar office buildings in the vicinity that are let at market rents have been sold recently on an 8% basis.

Calculate:

(a) the rent to be paid on the surrender of this lease for a new 25-year lease with 5-year rent reviews.

(a) the price the landlord should pay the tenant to secure a new 25-year lease to the tenant at MR.

## Marriage or synergistic value

When a property is split into multiple interests, either physically, or legally, or both, each of the newly created interests will have a market value. The total of these values will not necessarily equate with the market value of the freehold in possession of the whole property. In such cases, an element of what is known as 'marriage value' might be shown to exist. The RICS now refer to this as synergistic value (2011 VS Glossary).

The following example will serve to demonstrate this.

---

**Example 8.7**

A is the freeholder of an office block, the market rent of which is £28,000 pa on FRI terms.

Fourteen years ago A let the whole to B on a 40-year lease at a rent of £10,000 pa on FRI terms without a rent review.

B sub-let to C six years ago, at a rent of £18,000 pa FRI for a term of 25 years without rent review.

B wishes to become the freeholder in possession of the office block. Advise him on the sum to be offered for the interests of A and C.

How much should they accept?

**Valuation of current interests:**
(Freehold rate when let at MR with regular rent reviews 8%)

**1 A's interest:***

| | | |
|---|---|---|
| Rent | £10,000 | |
| PV £1 pa or 26 years at 10%* | 9.1609 | |
| | | £91,609 |
| Reversion to MR: | £28,000 | |
| PV £1 pa perp. defd. | | |
| 26 years at 8% | 1.6900 | £47,320 |
| Sum of term and reversion | | £138,929 |

*Valuers' adjustment to reflect the fixed nature of term income.

**2 B's interest:**

| | | |
|---|---|---|
| Term | | |
| Rent received: | £18,000 | |
| Rent paid: | £10,000 | |
| Profit rent: | £8,000 | |
| PV £1 pa for 19 years at 12% | 7.3658 | |
| | | £61,086 |
| Reversion to | | |
| MR | £28,000 | |
| Rent paid | £10,000 | |
| Profit rent | £18,000 | |
| PV £1 pa for 7 years at 12% | 4.5638 | |
| PV £1 in 19 years at 12% | 0.1161 | |
| | 0.5298 | £9,537 |
| Term plus reversion | | £70,623 |

**3 C's interest**

| | | |
|---|---|---|
| Rent received | £28,000 | |
| Rent paid | £18,000 | |
| Profit rent | £10,000 | |
| PV £1 pa for 19 years at 12% | 7.3658 | |
| Estimated value | | £73,658 |

The total value of all interests at present is £138,929 + £70,623 + £73,658, a total of £283,210.

B wishes to become freeholder in possession. How much will this be worth now that the MR can be received immediately and in perpetuity?

| | | |
|---|---|---|
| MR | £28,000 | |
| PV £1 pa in perp at 8% | 12.50 | £350,000 |

Notice that the total value of all interests at present is only £283,210, so there will be a marriage value, created by the merger of interests, of £66,790. In law when a lease is bought back by a freeholder, it is deemed to have been terminated by the merger of two interests in the same parcel of real estate; this means that the terms 'merger' and 'marriage' are frequently substituted for each other. Generally, the market value of a freehold in possession will be a more attractive investment than a freehold subject to a number of leases and sub-leases. The reality is that this could become a 7% investment boosting the marriage value by a further £50,000 (£28,000/0.07 = £400,000 which is £50,000 more than at 8%).

How is this marriage created, and where does it arise?

Capital value is the product of two things: income and a capitalisation factor. It follows that the marriage value must arise from one or both of these.

1. *Income*: Does this change upon merger of the interests? In the case of the freehold in possession, the total income passing is £28,000. When A, B and C have interests in the property A receives £10,000, B makes a profit rent of £8,000 and C makes a profit rent of £10,000. The total of rents and profit rents passing is therefore £28,000.

This holds for any year. The total of rents and profit rents will always be equal to market rent, because what the freeholder loses by way of rent someone else gains as profit rent.

2. *Capitalisation factor*: Does this change? The freehold in possession is valued by applying a single rate Years' Purchase (YP) to the whole £28,000. However, when the interests are split, the profit rents earned by B and C are valued at higher yields to reflect the market's concerns regarding the risks inherent in short leases.

This leads to the value of the sum of the parts being lower than the value of the freehold in possession. In addition the freehold encumbered, as it is, is less attractive than the unfettered market let freehold as so it is valued at a higher rate.

The effect of splitting the freehold in possession into three interests has been to reduce the total value of the block, owing to the effect of valuing the same total income in parts at different yield rates. When leasehold and freehold interests are merged to create a freehold in possession, the total value is increased, because the whole income is valued at the same lower rate. The scale of the marriage value will be due to the market's perception of the risks attaching to each marketable interest. Ability to support the yields used is therefore critical in any negotiations. The question as to whether marriage value is fact or fiction can only be answered in the market place.

This increase in value created by the merger is known as marriage value.

| | |
|---|---|
| Marriage value | £66,790 |
| C | £73,658 |
| B | £70,623 |
| A | £138,929 |
| Added together becomes | £350,000 |
| Unencumbered freehold: | £350,000 |

This means that in negotiations B can offer an inducement sum over and above the market value of A's and C's interests, because of the additional value due to the merger. The total additional sum is £66,790 but that would mean no participation in the marriage sum in respect of B's current interest. There is an argument for suggesting that the £66,790 should be apportioned pro rat between the three interests. This is very simplistic and ignores the vagaries of market value and of market negotiations.

But how much should B offer to A and to C?

What price should A and C ask for their interests?

B's first move will be to buy either A's or C's interest. The maximum B will be able to offer to A is the gain to be made from the transaction. If A's interest is purchased the freeholder will only be subject to C's under lease.

Value of B + A

| | | |
|---|---|---|
| Rent | £18,000 | |
| PV £1 pa for 19 years at 9%* | 8.9501 | £161,101 |
| | | |
| Reversion to | £28,000 | |
| PV £1 pa in perp defd | | |
| 19 years at 8% | 2.8964 | £81,099 |
| Total | | £242,200 |

*+1% for long lease without review.

B's interest is worth £70,623; the gain will be £242,200 − £70,623 = £171,577 and this is the maximum that can be offered to A. The minimum that A might accept is the market value of £138,929. Assuming that A and B will employ valuers who will be aware of both figures, agreement will be reached between these two boundaries.

The difference between these figures is:

£171,577 − £138,929 = £32,648 and is the marriage value between A and B.

It can also be found in the following way.

A, before the transaction, had an interest worth £138,929; B had an interest worth £70,623; this totals £209,552. The value of the combined interest is £242,200; and the difference between these two figures is £32,648, the marriage value.

B's next step will be to acquire the interest of C. It must be noted that B now has an interest, as freeholder, that is worth £242,200. By acquiring C's interest, B becomes the freeholder in possession with an interest valued at £350,000. The gain that B stands to make is therefore:

£350,000 − £242,200 = £107,800

This is therefore the maximum that B could offer to C.

The minimum that C will accept will be the market value of £73,658. There will again be negotiation between these two figures. The difference between these two figures is £34,142, i.e. the marriage value between B and C.

This may also be obtained by summating the present interest of B and C (£242,200 and £73,658 = £315,858) and deducting this from B's new interest worth £350,000.

£350,000 − £315,858 = £34,142.

Note that the total marriage value was found to be £66,790.

This is split between A and B and B and C:

| | |
|---|---|
| Marriage value A + B: | £32,648 |
| Marriage value B + C: | £34,142 |
| Total marriage value: | £66,790 |

This term 'marriage value' usually refers to the above, the result of the merging of interests in the same property. It can also be used to describe the extra value created by a merger of two properties.

---

**Example 8.8**

B is the owner of a derelict house on a small site. On its own, the site is too small to be profitably developed, but could be used as a parking space. A is in a similar position next door. The area is zoned for shopping use and an indication has been given that planning permission would be given to construct a small shop covering both sites.

Assess the price that B could offer to A for the freehold interest in this property and a reasonable sale price.

Value of B's present interest:

| | |
|---|---|
| Rent for parking, say | £1,000 pa |
| PV £1 pa in perp at 10% | 10 |
| MV | £10,000 |

Value of proposed merged site.

This might be valued by using a 'residual valuation' (see Chapter 10) but market evidence indicates that similar sites have been selling for £30,000.

The maximum B can offer will therefore be £20,000 (£30,000 − £10,000). The minimum A will accept is £10,000 and a price will be agreed between these two figures through negotiation. A marriage value of £10,000 has been created (two plots at £10,000 is £20,000 but merged together make £30,000.

Throughout this chapter conventional capitalisation approaches to valuation have been followed. Elsewhere, a number of these approaches have been questioned. It is therefore necessary to emphasise that it is the valuer's role to assess market value. If the valuer feels confident that the particular approach adopted can be substantiated, by reference to the market, then marriage value in multiple interest investments may well exist; but the valuer must be certain of this market fact. The valuer must be certain that it is not based on false assumptions and fortuitous mathematics.

For example, an approach adopted in the United States for valuing leasehold investments in property could be loosely described as the 'difference in value' method. The logic used is that if a property has a rental value of £28,000 a year, and on that basis has a market value of £350,000, but due to the grant of certain leases is currently worth only £138,929 then the value of the leasehold investments is £350,000 less £138,929. On this basis marriage value does not exist. The most probable situations where marriage value will be found include: ransom strips in development valuations where one landowner controls access and can ransom the developer for a full marriage value contribution on top of market value and where a lease with a long unexpired term at a low rent is to be merged with a freehold.

This, then, is one extreme. At the other, consider the value of a very short leasehold interest. The value is likely to be low owing to the short term and the problem of dilapidations. The freehold interest would, however, reflect fully the loss of rental until review and in such a case genuine marriage value would exist.

It should be emphasised that the single rate valuation of leasehold interests has been adopted in these examples; conventional dual rate valuation of leaseholds (see Appendix A) may well accentuate the size of the marriage value, especially where a net of tax approach is adopted. For clarity it should be noted that in the case of the merger of interests, any concerns about recovery of capital can be resolved, because freehold and leasehold are merged and so capital can be recovered by the sale of the freehold. No worries over onerous tax issues will arise because all short-term investment elements are now merged into the freehold. However, as demonstrated in Chapter 6, it may be necessary to use DCF to fully reflect the future benefits of the leasehold interest.

A discussion of marriage value could not be complete without some mention of 'break-up' values. This in essence is the opposite of marriage value and recognises that different investors have different needs, objectives, risk preferences and tax positions. As one might purchase a company as a whole and then sell off the component parts separately to realise a higher total value, so one might buy a freehold interest in a property and through careful sub-division of title realise a higher total value. The idea is really no more complex than that of letting a large estate. This again emphasises the need to couple a thorough knowledge of the discounting technique with a thorough knowledge of the property market. A particular issue here is the whole question of unitisation of single property investments. The market will dictate whether 1,000 units in a £100 million property will be worth £100,000 per unit or more.

The test is not 'do the calculations suggest that marriage value exists?', but 'does the market evidence prove that marriage value exists?'

## Market rent, non-standard reviews, constant rent theory

How much rent should be paid under a lease with non-standard rent review periods so as to leave the two parties in the same financial position as if they had agreed on a standard term?

In times of rental increases, or, indeed, decreases, the rent review pattern will affect the full market rent of a property, a concept so far regarded in this book as being inflexible. It will be to the landlord's advantage in times of rental growth to insist upon regular rent reviews. Conversely, in periods of rental decline a longer fixed rental period might be preferred. The initial rent-based on market assumptions of five year, upward only rent reviews on FRI terms, with rent payable quarterly in advance - may need reconsideration where the proposed terms differ. Indeed it has already been shown in Chapter 4 how an FRI rent will need to be adjusted if the proposed terms are to be internal repairing with insurance recovery.

Two typical situations arise where such adjustments may be applicable. First, a new lease may be arranged without the prevalent rent review pattern. For example, in a market where three-yearly reviews are normally accepted, a new lease may be granted with seven-yearly reviews. In such a case the landlord might ask for a higher initial rent.

Second, the problem might arise at a rent review where the period between reviews is the result of previous negotiations. For example, a 42-year lease granted 21 years ago with a single midterm review will present problems if three-yearly reviews are currently accepted, and again the landlord might have to ask for a higher reviewed rent as compensation. A range of other flexible lease terms may need to be considered, the fundamental underlying rule in such calculations is that of maintaining value equivalence.

Several methods have been devised to deal with this problem in a logical manner. One method was illustrated in a letter by the late Jack Rose to the *Estates Gazette*, 3 March 1979, at page 824.

This method is designed to produce a single factor, k, which is applied to the market rent on the usual pattern to arrive at the adjusted rental value. This is formulated as:

$$k = \frac{(1+r)^n - (1+g)^n}{(1+r)^n - 1} \times \frac{(1+r)^t - 1}{(1+r)^t - (1+g)^t}$$

where $k$ = multiplier
$r$ = equated yield
$n$ = number of years to review in subject lease
$g$ = annual rental growth expected
$t$ = number of years to review normally agreed.

**Example 8.9**

What rent should be fixed at a rent review for the remaining seven years of a 21-year lease of shop property, when rents are expected to rise at 2% per annum and three-yearly reviews are prevalent in the market?

An equated yield of 10% based on the return provided by undated government stock with an adjustment for risk should be used, and the MR with three-yearly reviews would be £15,000 pa.

$$k = \frac{(1+0.10)^7 - (1+0.02)^7}{(1+0.10)^7 - 1} \times \frac{(1+0.10)^3 - 1}{(1+0.10)^3 - (1+0.02)^3}$$

$$= \frac{1.9487 - 1.1487}{0.9487} \times \frac{0.3310}{1.3310 - 1.0612}$$

$$= 0.8433 \times 1.2268 = 1.0346$$

Rent to be charged:
£15,000 x 1.0346 = £15,519 pa

There is growing evidence in the property market that the rent review pattern in a lease has a considerable effect upon the rents required by landlords, and the above is an attempt to make such adjustments logically. This method and the others suggested are adaptations of DCF techniques to valuation problems. It must be concluded that the use of DCF in all valuations will remove all of the inconsistencies referred to in this chapter, as well as those problems identified in Chapters 5, 6 and 7, and is the ultimate and logical refinement of 'the income approach'.

**Question 3: Non standard rent review**

Rents on five year rent review patterns are £25,000 a year. What rent should a landlord expect if the property is to be let with rent reviews every three years? Use a growth rate of 5%, and an equated yield of 10%.

The logic behind the DCF approach is difficult to defeat, but the market is reluctant to move into the uncharted area of market forecasting. The valuer is trying to establish, on the basis of the market rent for a specific fixed term, what the rent should be for a longer or shorter fixed term. The first is assumed to equate with the actual annual rentals each year for the term discounted to their present worth and reassessed as a fixed annual equivalent sum. The second is the projection or shortening of the actual expected annual sums re-expressed as an annual equivalent. This process can only be undertaken on the basis of an assumed (but hopefully research-supported growth rate) or on the basis of an implied growth rate.

The latter approach is more questionable in this exercise, because the implied rental growth figures are long-term averages to be used for the purpose of valuing non-market rent review patterns. Either the landlord or the tenant would be concerned if the actual rental value rose at a higher or lower rate than the implied average growth rate.

For an uplift rent, the valuer is trying to determine the appropriate growth rate for that specific property. There is then an argument in favour of using forecasted growth rates based on thorough research for assessing uplift rents rather than merely using implied rates.

In arbitration cases it is a question of presenting the strongest supportable case for or against the uplift; the arithmetical assessment is only a starting point.

In the case of non-normal rent reviews, the courts have adopted a simple 10 to 15% increase for a longer review, or increased the market rent by a factor 1.5% to 2% for each additional year over the normal rent review evidence in preference to a DCF approach.

The examples in this chapter do not reflect the statutory rights of landlords and tenants, which may have to be taken into account and are considered in Chapter 9.

## Quarterly to monthly

The structure of leases is constantly changing as the nature of business needs change. In the 1950s, leases of 50 years with rent reviews after 25 years were not uncommon; the emergence of inflation caused the market to move to 21-year rent reviews, then to seven and now five or three years. However, in almost all leases the rent was payable quarterly in advance. Currently leases are much shorter, often less than 10 years long and in many locations three years is popular. Now there is increased demand from tenants to pay their rent monthly in advance. This eases their cash flow and provides an earlier alert to landlords of tenant financial difficulties. The question here is what rent should be agreed on a monthly basis if the market evidence is quarterly?

From a landlord's point of view, one would expect it to be marginally greater when expressed as an annual rent. From a financial perspective, three month's rent in advance is better than one month's rent in advance. From a rent collection, accounting and management perspective, it will be greater because management costs are greater.

### Example 8.10

A property would let at a market rent of £20,000 a year with rent payable quarterly in advance. If the market yield for this type of property on typical lease terms in this location would be 8% what should be the annual rent where the payments are monthly in advance?

**Step 1** A true equivalent yield (TEY) must be calculated. The TEY is the yield reflecting the frequency of payment, in this case four times a year.

$$r = \left[ \frac{1}{\left(1 - \dfrac{i}{4}\right)^4} - 1 \right]$$

Substituting 0.08 (8%) for $i$ and solving to find r (the TEY) the equation becomes:

$$r = \left[ \frac{1}{(1-0.02)^4} - 1 \right] = \left[ \frac{1}{0.922368} - 1 \right] = 1.084166 - 1 = 0.084166$$

In other words the TEY with quarterly payments is not 8% but 0.084166 × 100 which is 8.4166%.

**Step 2** Convert the TEY to a monthly yield. If $(1+i)^n = 1.084166$ then to find the monthly yield

$$i = (\sqrt[12]{1.084166}) - 1 \; ; i = (1.0067569) - 1 \; and \; \therefore$$
$$i = 0.0067569 \times 100 = 0.67569 \% \; per \; month$$

**Step 3** A £20,000 rent at 8% gives market value of £250,000. The 8% is nominal and as it is let quarterly in advance. The TEY is 8.4166%. The landlord, in agreeing to move to monthly in advance, will not wish to lose market value and so will need more than £20,000, i.e. more than £1,666.67 per month, if not then the TEY of 8.4166% will not be maintained.

And so $\left[ \left( \frac{1}{0.0067569} \right) + 1 \right] \div 12 = 12.4164 \; and \; \dfrac{£250,000}{12.4164} = £20,134.$

At this annual rent the value is maintained if rent is paid monthly in advance. However, this will not be the end of the negotiations as monthly rental investments might be regarded as poorer by the market and yields may move out to say 8.20% or whatever is sufficient to compensate for the more regular but smaller payments. The landlord might also need an uplift to cover the extra accounting costs associated with monthly banking. Hence, a rule of thumb might be to add 1% or 2% to the normal market rent when moving from quarterly to monthly. Here this would be between £20,200 and £20,400.

(This approach is based on the teaching of J Armatys at Sheffield Hallam University but any errors in interpretation are the authors'.)

## Summary

- Premiums are payable by tenants to landlords to ensure a letting at below market rental or for other reasons.
- Reverse premiums are payable by landlords to tenants or by tenants to sub-tenants to secure deals above market rentals.
- Premiums are generally assessed on the basis of value equations. Both parties should be neither better off nor worse off, and a before and after valuation is preferred.
- At the end of a lease either landlord or tenant may seek to secure a new lease in exchange for the surrender of the old one. Neither party should be any better off nor any worse off following the surrender and renewal, and a present vs. proposed interest calculation is normally used.

- Where multiple interests exist in a single property, marriage value may exist.
- Constant rent theory argues that a landlord should be no worse off over time by letting on an extended rent review pattern, but the market places an upper limit of 10% to 15% on the uplift negotiable for longer intervals between rent reviews.
- A similar issue has emerged with the move from upward only rent reviews to upward/downward rent reviews. Here again one can establish through the use of DCF techniques the amount of additional rent a landlord would require to compensate for the loss of security attaching to upward only rent reviews, but tenants may not be willing to meet such uplifts from year one. Tenants would argue that if rents do rise then they may well end up paying twice; once through the uplift and once through the increased rent at the rent review.
- When using quarterly market rent data for a lease with monthly payments an uplift may be required by the landlord; similar equivalent calculations may be necessary to arrive at a revised annual figure but a rule of thumb approach is likely to prevail in the market.

# Chapter 9

## The Effects of Legislation

## Introduction

In the examples so far it has been assumed that landlords and tenants are free to negotiate whatever lease terms they find acceptable for a particular tenancy of a property. In particular, it has been assumed that they are free to agree the amount of rent to be charged for the premises and that in most cases the landlord can obtain vacant possession at the end of the current lease. This is not always the case in the United Kingdom and the valuer must have regard to the provisions of the Landlord and Tenant Acts and the various Rent and Housing Acts as they affect properties occupied for business and residential purposes. The Agricultural Holdings Acts are of similar importance in the case of agricultural property, but agricultural property is not considered in this text.

The predominant feature of all landlord and tenant legislation is that it amends the normal law of contract as between landlord and tenant, giving tenants substantial security of tenure (the right to remain in occupation following the termination of a contractual agreement) and setting out statutory procedures for determining and sometimes controlling the rents that a landlord can charge a tenant for the right to occupy a property. (The subject matter of this chapter is discussed in more detail in Baum et al (2007).)

## Business premises

The most important statutes affecting business premises are the Landlord and Tenant Act 1927 Part I, and the Landlord and Tenant Act 1954 Part II, as amended by the Law of Property Act 1969 Part I, and by the Regulatory Reform (Business Tenancies) (England & Wales) Order 2003.

These statutes primarily affect industrial and commercial property, but the expression used is 'business includes a trade, profession or employment and includes any activity carried on by a body of persons whether corporate or not'. This definition as set out in section 32 of the Landlord and Tenant Act 1954 Part II is sufficiently broad to include some types of occupation which would not normally be regarded as business occupations, such as a tennis club (see *Addiscombe Garden Estates Ltd* v *Crabbe* (1958) 1 QB 513).

## Compensation for improvements

Under section 1 of the Landlord and Tenant Act 1927 Part II, the tenant of business premises is entitled at the termination of the tenancy, on quitting his holding, to be paid by his landlord compensation in respect of any improvement (including the erection of any building) on his holding made by him or his predecessors in title, not being a trade or other fixture which the tenant is by law entitled to remove, which at the termination of the tenancy adds to the letting value of the holding.

This right does not extend to improvements carried out before 25 March 1928, nor to improvements, according to section 48 of the Landlord and Tenant Act 1954,

> made in pursuance of a statutory obligation, nor to improvements which the tenant or his predecessors in title were under an obligation to make, such as would be the case where a tenant covenanted to carry out improvements as a condition of the lease when entered into

Except that those made after the passing of the 1954 Act 'in pursuance of a statutory obligation' will qualify for compensation (section 48 of the Landlord and Tenant Act 1954).

The tenant will normally require the consent of the landlord before carrying out alterations or improvements, or alternatively they may apply to the court for a certificate to the effect that the improvement is a 'proper improvement' (section 3 of the Landlord and Tenant Act 1927).

It should be noted under section 19(2) of the Landlord and Tenant Act 1927 Part II that

> in all leases...containing a covenant, condition or agreement, against the making of improvements without licence or consent, such covenant...shall be deemed...to be subject to a proviso that such licence or consent is not to be unreasonably withheld...

Section 49 of the 1954 Act renders void any agreement to contract out of the 1927 Act.

Compensation payable is limited under Schedule 1 of the Landlord and Tenant Act 1927 to the lesser of:

1. The net addition to the value of the holding as a whole as a result of the improvement.
2. The reasonable cost of carrying out the improvement at the termination of the tenancy, subject to a deduction of an amount equal to the cost (if any) of putting the works constituting the improvement into a reasonable state of repair, except as so far as such cost is covered by the tenant's repairing covenant.

Further, 'in determining the amount of such net addition regard shall be had to the purposes for which it is intended the premises shall be used after the termination of the tenancy' (section 1(2) of the Landlord and Tenant Act 1927); for example, if the premises are to be demolished immediately then the improvements are of no value to the landlord and no compensation would be payable. However, if the premises are to be demolished in, say, six months time and there is a temporary user planned, then compensation would be based on the net addition to value of the improvements for that six-month period. If the landlord and tenant fail to agree as to the amount of compensation, the matter can be referred to the county court. Where a new lease is granted on the termination of the current lease, no compensation can be claimed at that point in time. Both the 1927 Act and the 1954 Act provide that the rent on a new lease shall exclude any amount attributable to the improvements in respect of which compensation would have been payable.

A valuer, when instructed to value business premises, must consider the extent to which they have been improved by the tenant.

---

**Example 9.1**

Value the freehold interest in office premises currently let at £50,000 a year on full repairing and insuring (FRI) terms with six years of the lease unexpired. The current market rent (MR) on FRI terms is £100,000 a year. Improvements were carried out by the tenant and these have increased the MR by £10,000 and would cost today an estimated £80,000 to complete.

Value the premises on the assumption that the tenant is vacating at the end of the present lease and will be able to make a valid claim for compensation under the Landlord and Tenant Act 1927.

| | | |
|---|---|---|
| Current net income | £50,000 | |
| PV £1 pa for 6 years at 8% | 4.62 | £231,000 |
| Reversion in 6 years to | £100,000 | |
| PV £1 pa in perp def'd 6 years at 8% | 7.87 | £787,000 |
| | | £1,018,000 |

Less compensation under Landlord and Tenant Act 1927 for improvements being the lesser of:
(a) Net addition to the value
Increase in rent attributable to improvements:

| | | |
|---|---|---|
| | £10,000 | |
| PV £1 pa in perp at 8% | 12.5 | |
| | £125,000 | |

(b) Cost of carrying out improvements
The cost is £80,000 and therefore compensation is £80,000 because it is the lower amount. This is the amount it would cost today but it will not have to be paid until the end of the lease and so in value terms there is an argument for discounting this for six years to its present value:

| | | |
|---|---|---|
| | £80,000 | |
| PV £1 for 6 years at 8% | 0.63 | |
| | £50,400 | |

Therefore, the market value of the freehold allowing for payment of compensation to the tenant is £967,600 (£1,018,000 − £50,400 = £967,600) say £967,000.

As shown elsewhere, this traditional, equivalent yield valuation raises some basic issues. The first is the whole question of implicit versus explicit growth valuation models, and the second is the validity of using current estimates of changes in rent attributable to the improvements and of using current costs for assessing the compensations when the actual compensation will have to be based on figures applicable in six years time. This is a further argument for at least checking the valuation on a discounted cash flow basis (DCF) basis.

It could be argued that the £80,000 should not be discounted as in effect it is the equivalent cost of the future amount discounted; in which case a prudent purchaser would only offer £938,000 (£1,018,000 – £80,000).

Whilst valuers may be asked to assess the amount of compensation to be paid under a 1927 claim, it is rare for market values to be affected by the future possibility of such payments. Such a valuation is only likely to occur when a tenant who has undertaken improvements has served notice of his or her intention to vacate.

## Security of tenure

In accordance with the provisions of section 24(1) of the Landlord and Tenant Act 1954, tenants of business premises are granted security of tenure; however, the parties may contract out of these provisions. Tenancies to which this part of the Act applies will not come to an end unless terminated in accordance with the provisions of Part II of the Act, so that some positive act by the landlord or tenant needs to be taken to terminate a tenancy. Where notice to terminate is served by the landlord, it must be for at least six months and for not more than 12 months. Such notice cannot come into force before the expiration of an existing contractual tenancy. Thus, in the case of most leases of business premises, the earliest date which a landlord can serve notice on a tenant is 12 months prior to the contractual termination date. If notice is not served, the tenancy continues as a statutory tenancy at the contracted rent until terminated by notice.

The Act further provides that a tenant has the right to the renewal of their lease. If the landlord wishes to obtain possession they may oppose the tenant's request for a new tenancy only on the grounds set out in the Act.

Section 30(1) states the following grounds on which a landlord may oppose an application:

(a) where under the current tenancy the tenant has any obligations as respects the repair and maintenance of the holding that the tenant ought not to be granted a new tenancy in view of the state of repair of the holding, being a state resulting from the tenant's failure to comply with the said obligations;

(b) that the tenant ought not to be granted a new tenancy in view of his persistent delay in paying rent which has become due;

(c) that the tenant ought not to be granted a new tenancy in view of other substantial breaches by him of his obligations under the current tenancy, or for any other reason connected with the tenant's use or management of the holding;

(d) that the landlord has offered and is willing to provide or secure the provision of alternative accommodation for the tenant, that the terms on which the alternative accommodation is available are reasonable having regard to the terms of the current tenancy and to all other relevant circumstances, and that the accommodation at the time at which it will be available is suitable for the tenant's requirements, including the requirement to preserve goodwill, having regard to the nature and class of his business and to the situation and extent of and facilities afforded by the holding;

(e) Where the current tenancy was created by the sub-letting of part only of the property comprised in a superior tenancy and the landlord is the owner of an interest in reversion expectant on the termination of that superior tenancy, that the estimate of the rents reasonably obtainable on separate lettings of the holding and the remainder of that property would be substantially less than the rent reasonably obtainable on a letting of that property as a whole, that on the termination of the current tenancy the land-lord requires possession of the holding for the purpose of letting or otherwise disposing of the said property as a whole, and that in view thereof the tenant ought not to be granted a new tenancy;

(f) That on the termination of the current tenancy the landlord intends to demolish or reconstruct the premises comprising the holding or a substantial part of those premises or to carry out substantial work of construction on the holding or part thereof, and that he could not reasonably do so without obtaining pos-session of the holding;

(g) Subject as hereinafter provided that on the termination of the current tenancy the landlord intends to occupy the holding for the purposes, or partly for the purposes, of a business to be car-ried on by him therein, or as his residence.

To oppose an application under ground (g), the landlord must have been the owner of the said interest for at least five years prior to the termination of the current tenancy. Section 6 of the Law of Property Act 1969 extends section 30(g) of the 1954 Act to companies con-trolled by the landlord and section 7 of the Law of Property Act 1969 has altered the effects of 30(f) of the 1954 Act so that a landlord wish-ing to oppose the grant of a new tenancy under that ground must now not only prove an intent to carry out substantial works of alter-ation, but also that it is necessary to obtain possession in order to complete such works. Thus, if a landlord can demolish and rebuild without obtaining possession, and if the tenant is agreeable or will-ing to co-operate then the courts will allow the tenant to remain in possession.

Section 32 of the Landlord and Tenant Act 1954, whilst requiring the new lease to be in respect of the whole of the building, has now been amended by section 7 of the Law of Property Act 1969, which adds sections 31(A) and 32(1)(A) to allow a court to grant a new tenancy in respect of part of the original holding where the tenant is in agreement.

## Compensation for loss of security

When a landlord is successful in obtaining possession the tenant may be entitled to compensation under section 37 of the Landlord and Tenant Act 1954.

> Where the Court is precluded…from making an order for the grant of a new tenancy by reason of any of the grounds specified in paragraphs (e), (f) and (g)… the tenant shall be entitled on quitting the holding to recover from the landlord by way of compensation an amount determined in accordance with the following provisions of this section…

The amount of compensation payable will be two times the rateable value of the holding if, for the whole of the 14 years immediately preceding the termination of the tenancy, the premises have been occupied for the purposes of a business carried on by the occupier, or if during those 14 years there had been a change in the occupation and the current occupier is the successor to the business carried on by a predecessor. In all other cases the amount of compensation is the amount equal to the rateable value of the holding as at the date of assessing the compensation.

Additional compensation is payable under section 37A. This is payable if a landlord obtains possession or is successful in opposing a tenant's request for a new lease and is shown subsequently to have succeeded due to misrepresentation or has concealed material facts. The amount in these cases is determined by the court as 'such sum as appears sufficient as compensation for damage or loss sustained by the tenant…' Assessment of such a claim is likely to be based on factors other than property factors, such as loss of trade and profits.

## Terms of the new lease

Where a new lease is granted then the new rent payable will normally be in accordance with the provisions of section 34 of the Landlord and Tenant Act 1954, particularly when the parties are in disagreement and the matter is referred to the county court for settlement.

The Law of Property Act 1969 has amended section 34 so that:

> the rent payable under a tenancy granted by order of the Court under this part of this act shall be such as may be agreed between the landlord and tenant or as, in default of such agreement, may be determined by the court to be that at which, having regard to the terms of the tenancy (other than those relating to rent), the holding might reasonably be expected to be let in the open market by a willing lessor, there being disregarded:
>
> (a) any effect on rent of the fact that the tenant has or his predecessors in title have been in occupation of the holding;
> (b) any goodwill attached to the holding by reason of the carrying on thereat of the business of the tenant (whether by him or by a predecessor of his in that business);

(c) any effect on rent of any improvement carried out by the tenant or predecessor in title of his otherwise than in pursuance of an obligation to his immediate landlord (see LPA amendment below);

(d) in the case of a holding comprising licensed premises any addition to its value attributable to the licence, if it appears to the court that having regard to the terms of the current tenancy and any other relevant circumstances the benefit of the licence belongs to the tenant.

Items (a),(b) and (c) are those that valuers have most frequently to reflect in valuations of business premises.

Items (a) and (b) cause particular difficulty in assessment because, whilst it is simple to explain their meaning, it is more difficult to assess their impact in value terms. Under item (a), the valuer must demonstrate that, for example, if the occupying tenant would bid £55,000 but the premises have a rent as defined in the Act of only £50,000, then only £50,000 is payable. The potential tenant overbid of £5,000 in a free market must be disregarded.

Similarly under item (b) if it can be demonstrated that the premises are worth £50,000 but to any other tenant carrying on the same business are worth £55,000 then the £5,000 of business goodwill must be disregarded. This is because in both cases it is this tenant and their business that has pushed rental bids up.

Section 1 of the Law of Property Act 1969 has extended the meaning of section 34(c) to include tenants' improvements carried out at any time within 21 years of the renewal of the tenancy.

All the other terms and conditions of a new tenancy shall be as agreed between the parties, but if the parties cannot agree then sections 33 and 34 of the Landlord and Tenant Act 1954 require the court to restrict the terms of the tenancy to 15 years with appropriate rent reviews (Law of Property Act 1969 section 2 which adds subsection 3 to section 34 of the 1954 Act).

A number of Law Commission recommendations were implemented by the 2003 Order. The most significant from a valuation perspective was the increase to the maximum length of lease which can be granted by the courts from 14 to 15 years.

---

### Example 9.2

Assuming the facts as stated in Example 9.1, value the freehold interest and assume the tenant is granted a new 15-year lease with rent reviews in the fifth and 10th year and that the improvements were completed three years ago.

The first step to resolve this problem requires the sorting out of the income flow. In six years time the lease is due for renewal, at a rent ignoring the effect on rent of the improvements (Landlord and Tenant Act 1954).

However, there is some doubt as to the rent that could be charged at the rent reviews after five and 10 years. According to the Law of Property Act 1969, as amended, it would seem that no account should be taken of the value of the improvements for a period of 21 years.

If this argument applies then the rent on the reviews must once more be at a figure excluding any value attributable to the improvements, provided a section 34 disregard or equivalent is included in the rent review clause. This reasoning, coupled with the specific provisions in the Landlord and Tenant Act 1954, would effectively result in no account being taken of the improvements at any time during the whole of the new lease.

The income position for valuation purposes is as scheduled below:

| Time | Activity | Rent |
|---|---|---|
| −3 Years | Completion of improvements | |
| Today's date | Date of valuation | £50,000 |
| +6 years | Lease renewal disregarding effect of improvements | £90,000 |
| +11 years | First 5 year review | £90,000 |
| +16 years | Second 5 year review | £90,000 |
| +18 years | Improvements 21 years old | |
| +21 years | Next lease renewal | £100,000 |

Solution, assuming cash flow as shown above (and see notes following the valuation)

| | | | |
|---|---|---|---|
| Current net income | | £50,000 | |
| PV £1 pa for 6 years at 8% | | 4.62 | £231,000 |
| First reversion to S.34 rent | | £90,000 | |
| PV £1 pa for 15 years at 8% | 8.5595 | | |
| PV £1 in 6 years at 8% | 0.63017 | 5.3939 | £485,455 |
| Second reversion to MR | | £100,000 | |
| PV £1 pa in perp def'd 21 years at 8% | | 2.4832 | £248,320 |
| | | | £964,775 |

Notes:

1. No specific allowance has been made for compensation for improvements as the valuation assumes further renewals under Landlord and Tenant Act to the current tenant or successor in title.
2. Conventional equivalent yield valuation assumptions underlie the valuation, the review pattern is such that an equated yield or DCF explicit valuation should be used as a check.
3. The valuer can only make reasonable assumptions. To assume for valuation purposes that the tenant will forget they have rights under the1954 Act would be inferring a definition of market value that excluded acting prudently and with knowledge. To assume that the tenant's solicitors will forget to include a section 34 disregard in the rent review clause would be a false assumption and could give rise to an action for negligence against the valuer. The assumptions made in this solution are realistic and reasonable.

If a 21-year period elapses prior to a rent review date then the market rent (MR) of the improved premises could be charged from the review date.This would appear to be contrary to the 1954 Act as originally drafted, which clearly intended that the effects on rent of any improvements should be disregarded for the whole of the new tenancy, in this case for the whole of the 15 years.

Whilst some confusion apparently exists, a number of points are becoming obvious from the decisions reached in a number of landlord and tenant cases.

First, the legislation relates quite clearly to the determination of 'rent payable under a tenancy granted by order of the Court', and this can only occur on renewal of a lease and not on a rent review.

Second, the Act uses the word 'reasonable' which suggests that the rent as determined need not be the maximum possible rent.

Third, the rent on any review will be determined in accordance with the appropriate clauses in the lease and, unless there is specific reference to section 34 of the Landlord and Tenant Act 1954 or a specific statement that improvements carried out during an immediately preceding lease, or within 21 years of the review date, are to be disregarded, the review rent may fully reflect the current market rent of the property as improved.

Professional advisors are therefore forewarned when acting for tenants to see that rent review clauses in leases are sufficiently worded to protect their clients. In this example, if the tenant was to accept as the terms of the new lease a rent review clause which contained no disregards then the MR would be charged at the first rent review.

When licences for improvements are granted they should also confirm that the effect on rent of the improvements will be disregarded on review or renewal of the current lease. If this advice is followed then tenants may avoid any repetition of *Ponsford* v *HMS Aerosols Ltd* [1977] 2 EGLR 68, where tenants of a factory rebuilt the premises which had been destroyed by fire, at the same time substantially improving the property with the landlord's consent. Shortly after rebuilding, the rent was due for review.

The wording of the lease and licence was such that the Court of Appeal held that the tenants would have to pay rent in respect of the improvements, the cost of which they had borne themselves. The wording of the lease, together with the wording of any licences, will determine the factors to be taken into account when assessing the rent to be paid on review. A similar situation would occur if a tenant were to carry out improvements without the proper licence of the landlord. (Readers are also referred to *English Exporters (London) Ltd* v *Eldonwall Ltd* (1973) Ch 415.)

When valuing business premises, or advising tenants of business premises, the valuer must have regard to all the terms and conditions of the lease and to the relevant statutory provisions.

In practice, negotiations and or court proceedings may result in the new rent commencing many months after the termination of a current contractual lease. Section 64 of the 1954 Act further provides that any new terms, including those relating to rent, may only commence three months after the court application has been 'finally disposed of'. This can lead to a considerable loss of income for the landlord.

The Law of Property Act 1969 added a new section, 24A, to the 1954 Act.

**(1)** The landlord of a tenancy to which this part of this Act applies may:

**(a)** if he has given notice under section 25 of this Act to terminate the tenancy; or

**(b)** if the tenant has made a request for a new tenancy in accordance with Section 26 of this Act;

**(c)** apply to the court to determine a rent which it would be reasonable for the tenant to pay while the tenancy continues by virtue of Section 24 of this Act, and the court may determine a rent accordingly.

**(2)** A rent determined in proceedings under this section shall be deemed to be the rent payable under the tenancy from the date on which the proceedings were commenced or the date specified in the landlord's notice or the tenant's request, whichever is the later.

The Regulatory Reform Order 2003 extended this right to tenants. Landlords are likely to exercise this right when rents have risen and tenants when rents have fallen. The interim rent will normally be the same as the rent determined for the new tenancy, which means that, for valuation purposes, a valuer can assume that the new rent at renewal begins at the date of contractual termination of the old lease. The real position from a property management perspective may be different and this rental position will normally only be achieved if the right notices are served at the right times, in accordance with the legislation as amended by the Order. There may still be a case for interim rents to be settled on the basis of a tenancy from year to year. This cushioned rent is only likely to occur where a landlord has applied for an interim rent and has opposed the grant of a new tenancy. A rent from year to year tends to be lower than a normal MR.

## Landlord and tenant negotiations

It is important to appreciate that whilst the Landlord and Tenant Acts are there to protect the tenant on termination of a lease, many tenants will seek to negotiate new leases before their current leases expire. The Landlord and Tenant Acts give tenants increased bargaining strength and full regard should be given to these statutes when advising a landlord or tenant on the terms for a surrender and renewal of a business lease.

### Example 9.3

A tenant occupies shop premises on a lease having two years to run at £60,000 a year net. Ten years ago the tenant substantially improved the property. The MR of the property today as originally demised would be £100,000, but as improved it is £140,000. The tenant wishes to surrender her present lease for a new 15-year lease with reviews every five years to MR; the landlord has agreed in principle and you have been appointed as independent valuer to assess a reasonable rent for the new lease. Freehold capitalisation rates are 7% and leaseholds are 10%.

As outlined in Chapter 8, this requires consideration from the tenant's and landlord's points of view on a 'before and after' basis.

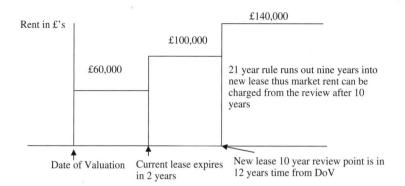

Rent in £'s

£140,000

£100,000

£60,000

21 year rule runs out nine years into new lease thus market rent can be charged from the review after 10 years

Date of Valuation    Current lease expires in 2 years

New lease 10 year review point is in 12 years time from DoV

Value of freeholder's present interest, i.e. assuming no surrender:

| | | | |
|---|---|---|---|
| Current income | | £60,000 | |
| PV £1 pa for 2 years at 7% | | 1.8080 | £108,480 |

| | | | |
|---|---|---|---|
| Reversion to MR subject to section 34(c) of the Landlord and Tenant Act 1954 Part II and assuming a new 15-year lease with review agreed | | £100,000 | |
| PV £1 pa for 10 years at 7% | 7.02 | | |
| PV £1 in 2 years at 7% | 0.87 | 6.10 | £610,000 |
| Reversion to MR | | £140,000 | |
| PV £1 pa in perp def'd 12 years at 7% | | 6.34 | £887,600 |
| Value of present interest | | | £1,606,080 |

(If the assumption here is that a review after five years would be permitted by the courts then one must further assume that professional advisors would see that the rent review clause fully reflected the intention of the Landlord and Tenant Act 1954 Part II, as amended by the Law of Property Act 1969.)

Value of freeholder's proposed interest:
Let rent to be reserved for new 15-year lease = £x a year

| | | |
|---|---|---|
| Proposed rent | £x | |
| PV £1 pa 5 years at 7% | 4.10 | 4.10x |
| Reversions to MR | £140,000 | |
| PV £1 pa in perp def'd 5 years at 7% | 10.185 | £1,425,900 |
| Value of proposed interest | | £1,425,900 + 4.10x |

On the assumption that the freeholder should be no better off and no worse off, the value of his present interest must be equated with the value of his future interest:

| | | |
|---|---|---|
| Present interest | = | Proposed interest |
| £1,606,080 | = | £1,425,900 + 4.1x |
| £1,606,080 − £1,425,900 | = | 4.1x |
| £180,180 | = | 4.1x |
| £43,940 | = | x |

Value of tenant's present interest assuming no surrender:

| | | |
|---|---|---|
| MR to tenant | £140,000 | |
| Less rent reserved | £60,000 | |
| Profit rent | £80,000 | |
| PV £1 pa for 2 years at 10% | 1.7355 | £138,840 |
| Reversion to | £140,000 | |
| Less section 34 rent reserved | £100,000 | |
| | £40,000 | |

| | | | |
|---|---|---|---|
| PV £1 pa for 10 years at 10% | 6.1446 | | |
| PV £1 in 2 years at 10% | 0.82645 | 5.078 | £203,128 |
| Value of present interest | | | £341,968 |

Note: After the rent review in 10 years the rent rises to £140,000.

Value of tenant's proposed interest:

Let rent to be reserved for new 15-year lease = £x per year, then the new profit rent is:

| | |
|---|---|
| MR to tenant | £140,000 |
| Less rent to be agreed | £x |
| Profit rent | £140,000 −£x |
| PV £1 pa for 5 years at10% | 3.7908 |
| Value of proposed interest | £530,712 − £3.7908x |

(Note both the £140,000 and the rent to be agreed of £x must be multiplied by the PV £1 pa.)

| | | |
|---|---|---|
| Present interest | = | proposed interest |
| £341,968 | = | £530,712-3.7908x |
| £3.7908x | = | £530,712-£341,968 |
| £3.7908x | = | £188,744 |
| £x | = | £49,790 |

Here it would seem reasonable for the parties to accept a rent for a new 15-year lease, on surrender of the present two-year term of, say, £45,000 a year with a MR review after five years. However, these valuations should be cross-checked with a DCF approach. A similar calculation using a dual rate of 8% adjusted for tax at 40% would produce a similar conclusion, although the leasehold figures would be different.

# Residential property

In this section the relevant Acts will be referred to by their initials and their year of enactment, e.g. Housing Acts will be HA 19yy, Rent Acts as RA 19yy, Landlord and Tenant Acts as LTA 19yy.

Over 68% (2001 Census) of homes in England and Wales are now owner-occupied and the appropriate method of valuation is direct capital comparison (see Mackmin (2008)). However, there are still many individuals and families occupying furnished and unfurnished property as tenants. The market value of such properties can be assessed by using the income approach or in appropriate cases, if vacant possession is possible, by comparison.

Considerable changes in this sector occurred in the 1980s. Many public sector tenants exercised their right to buy under the HA 1980 (subsequently HA 1985) and are now homeowners. HA 1988 made further changes to the law relating to public and private sector housing, including that applying to housing associations. From 15 January 1989 all existing and new assured tenancies come under the HA 1988 (amended by HA 1996). HA 1988 also introduced assured shorthold tenancies (ASTs) to replace the previous system of protected shorthold tenancies. Valuers dealing with residential investment valuation are referred to in www.odpm.gov.uk or www.dcl.gov.uk and those dealing with older tenancies, especially those created before 15 January 1989, are referred to in Baum (2007).

The HA 2004 increased local authority control over assessing housing conditions and enforcing housing standards, licensing houses in multiple occupation (HMOs) and other residential accommodation. Housing law is under continual review and valuers need to be vigilant when checking all residential tenanted properties; particular care is needed to check for improvement notices, prohibition orders or hazard awareness notices under HA 2004 Part I, HMO licences under HA 2004 Part II, selective licensing under HA 2004 Part III or other orders and notices under HA 2004 Parts I and IV, served by local authorities.

The Leasehold Reform Act 1967 (LRA 1967) introduced the right to enfranchise (the right to buy) or extend the lease of a house. The Leasehold Reform Housing and Urban Development Act 1993 (LRHUDA 1993) extended this concept to flats. The Commonhold and Leasehold Reform Act 2002 (CLRA 2002) has further amended the provisions of both the LRA 1967 and LRHUDA 1993, as well as introducing the new tenure system of commonhold. Leasehold reform is no longer considered in this book and readers are referred to Shapiro, Davies and Mackmin (2009) and Baum (2007).

## Private sector tenancies

Residential investment properties in the private sector can be grouped under the following headings:
1. Assured tenancies under the Housing Act 1988, as amended by HA 1996.
2. Assured shorthold tenancies under the HA 1988, as amended by the HA 1996.

The following residential investment properties are excluded from being either assured or assured shorthold tenancies:
- Tenancies subject to protection under RA 1977 Part I, generally referred to as regulated furnished or regulated unfurnished tenancies, including those tenancies previously known as controlled tenancies by section 64 of the HA 1980, except those to which LTA 1954 Part II applies. Tenancies in this category must have been 'entered into before, or pursuant to a contract of the HA 1988', i.e. 15 January 1989.

- Tenancies of high value dwelling-houses. These are (para 2 & 2A Schedule 1 of the HA 1988):
  - If tenancy or contract for tenancy was entered into before 1 April 1990, and the rateable value on 31 March 1990 was greater than £1,500 (Greater London) £750 (elsewhere).
  - If tenancy entered into after 1 April 1990, and the net rent is greater than £25,000 pa. This annual sum has now been raised to £100,000 pa (Assured Tenancies (amendment) Order 2010. This amendment came into force on 1 October 2010). It is retrospective and means that the majority of residential tenancies are now AST.
- Tenancies at a low rent (such low rent tenancies may be protected under the LTA 1954 Part I). These are (para 3 Schedule 1 of the HA 1988):
- If tenancy or contract for tenancy entered into before 1 April 1990, and the net rent is not greater than £1,000 (Greater London) £250 (elsewhere).
- If tenancy entered into after 1 April 1990 and with a rateable value as at March 1990 the net rent is less than two-thirds of the rateable value.
- Tenancies under LTA 1954 Part II (i.e. treated as business tenancies) and licensed premises (paras 4 and 5 Schedule 1 of the HA 1988).
- Agricultural land exceeding two acres let with dwelling-house or an agricultural holding under the Agricultural holdings Act 1986 (paras 6 and 7 Schedule 1 of the HA 1988).
- Lettings to students (para 8 Schedule 1 of the HA 1988).
- Holiday lets (para 9 Schedule a of the HA 1988).
- Tenancies where the landlord occupies as only or principal home at all times since a tenancy was granted of another part of the same building. Does not apply to purpose built block of flats unless landlord so resident in another part of the same flat (para 10 and Part III Schedule 1 of the HA 1988).
- Property owned by the Crown, local authorities and other statutory bodies (paras 11 and 12 Schedule 1 of the HA 1988).
- Lettings to asylum seekers under Immigration and Asylum Act 1999 Part VI.

In addition there are tenancies with enfranchisement or extension rights under the LRA 1967 (as amended) or LRHUDA 1993 (as amended). A further minor group of formerly controlled tenancies which are partially business lettings cannot become regulated tenancies (section 24(3) of the RA 1977) and fall to be considered under LTA 1954 Part II.

Each, other than leasehold enfranchisement, is considered below in terms of the valuation implications of the legislation relating to that category. Those involved with the letting and management of residential property are referred specifically to the above legislation and to the HA 1988 and the LTAs of 1985 and 1987.

## Assured tenancies

The HA 1980 introduced a new class of tenancy in the private sector known as the assured tenancy. This was intended to encourage the institutional investor to build new homes to let on the open market. Few tenancies were created under HA 1980, largely due to the restriction of the provisions to new dwellings provided by approved landlords.

The HA 1988 created a completely new scheme of assured tenancies, borrowing heavily from LTA 1954 and RA 1977. LTA 1954 is paralleled in terms of tenant protection, but grounds for possession are very similar to those found in RA 1977. These assured tenancies may be granted at market rents.

Under HA 1988, all tenancies of residential property created on or after 15 January 1989, other than those statutorily excepted by Schedule 1 of the HA 1988 (see above), were assured or ASTs. It should be noted that HA 1966 restricts, on a practical level, the creation of assured tenancies, with all tenancies of residential property created on or after 28 February 1977 (other than those statutorily excepted by Schedule 1 of the HA 1988) being ASTs, unless a notice is served on the tenant or included in the agreement that the tenancy is an assured tenancy not an AST.

Section 5 of the HA 1988 specifies that assured tenancies can only be brought to an end with a court order and to obtain such an order the landlord(s) must follow the procedures set out in the Act, and must specify the grounds for possession which must be one or more of those set out in Schedule 2 of the HA 1988, amended by HA 1966. In the case of grounds 1 to 8 inclusive, the court, if satisfied, must order possession; in all other cases the court may order possession.

The valuation of freehold interests in residential property subject to assured tenancies must be carried out by reference to comparable transactions. The comparables may need to be analysed with reference to either the income approach or the direct capital comparison approach or both. The valuer needs to be aware that the market may be using a net income approach or, in some places, a more unsophisticated gross income approach. In terms of direct capital comparison, the market sometimes values by using a percentage of vacant possession value.

## ASTs

Although HA 1980 introduced the concept of shorthold tenancies, it is the HA 1988 and HA 1996 that have developed the concept.

- From 28 February 1997, all new tenancies (except those in Schedule 1 of the HA 1988) will be ASTs, unless they are specified to be assured tenancies.
- Prior to 28 February 1997, an AST had to be for a minimum period of six months. From 28 February 1997, an AST can be for less than six months or on a periodic basis, but the tenant

has the right to stay for a minimum period of six months unless one or other of the grounds for possession set out in Schedule 2 of the HA 1988 is established and possession is ordered by the court.

- If the AST commenced prior to 28 February 1997, and either the fixed term has ended or the property is on a periodic tenancy, possession (without giving grounds) can be regained on two months written notice. If the AST commenced after 28 February 1997, possession may be regained on two months written notice after either the fixed term comes to an end, or at any time during a periodic tenancy, provided it is at least six months from the commencement of the original tenancy.

There are other (limited) provisions whereby possession may be obtained during a fixed term.

The ability to recover possession at the end of a fixed term or during a periodic tenancy suggests that such properties will normally be valued on the basis of capitalised term income plus a reversion to vacant possession market value or, since the period before obtaining possession can be relatively short, by direct reference to, or discount from, Vacant Possession (VP) values. On occasions there may be an obvious problem tenant who is not just going to vacate and in these circumstances the valuer needs to consider how long, and at what cost, possession might be delayed.

## Tenancies subject to the provisions of the Rent Act 1977

The protected or regulated tenancy is one subject to the RA 1977. Its main features are security of tenure and the provision of a 'fair rent'. Security of tenure was achieved by ensuring that once the contractual term came to an end, the tenancy effectively continued on the same terms and conditions, but as a statutory periodic tenancy. The most important feature of the 'fair rent' is that the effect on rent of the scarcity of residential accommodation within an area must be ignored. Thus, a 'fair rent' will, in areas of undersupply, be lower than the rent one would expect if the premises were offered in the open market in the absence of the Rent Acts.

No new, regulated tenancies can be created after 15 January 1989. Whether or not a tenancy is regulated depends upon the date of creation and complex provisions relating to the rateable value (RV) of the property at the appropriate day, which is normally 23 March 1965. The client's solicitor should be asked to confirm the status of any residential tenants. The abolition of domestic rating in 1990 led to the need for substantial and complex rules governing the RV requirements. However, in brief, a tenancy which is entered into after 1 April 1990 is not protected if:

- The appropriate day fell on or after 1 April 1973 and the dwelling-house has an RV on that day exceeding £1,500 in London, £750 elsewhere.

- The appropriate day fell on or after 22 March 1973 but before 1 April 1973 and the house had an RV exceeding £800 in London or £600 elsewhere and 1 April had an RV exceeding £1,500 in London, £750 elsewhere.
- The appropriate day fell on or before 22 March 1973 and the house:
  - on the appropriate day had an RV exceeding in London £400, £200 elsewhere
  - on 22 March 1973 had an RV exceeding £600 in London and £300 elsewhere
  - on 1 April 1973 had an RV exceeding in £1,500 in London, £750 elsewhere.
- The rent is less than two-thirds of RV on the appropriate day.
- The appropriate day is before 22 March 1973 and the RV on the appropriate day is greater than £400/£200 and the rent is less than two-thirds RV on 22 March 1973.

As far as the valuation of properties subject to regulated tenancies is concerned the important points to note are as follows:

- The recoverable rent is restricted to the level of a 'fair rent'.
- The rent can only be increased after two years.
- There are complex rules relating to succession following the death of the original tenant (Schedule 1 of RA 1977).

The following are also excluded from protection: dwelling-houses let with other land: with a payment for board or services; to students; holiday lets; agricultural holdings; licensed premises; resident landlords; Crown, Local Authorities and certain other statutory bodies; when under LTA 1954 Part II; when an assured tenancy under section 56 of the HA 1980.

The RA 1965 introduced the concept of 'fair rent' and this is now consolidated in the RA 1977. The rules and regulations relating to fair rents are subject to change and to variation due to court precedent. Basically, a fair rent must have regard to the age, character, locality and state of repair of the dwelling-house and, if any furniture is provided for the use of the tenancy, the quantity, quality and condition of the furniture. However, in determining a fair rent the following must be disregarded: any disrepair or defect attributable to a failure by the tenant to comply with the terms of the tenancy; any improvement carried out by the tenant otherwise than in pursuance of the terms of the tenancy; and any improvement to the furniture made by the tenant under the regulated tenancy or as the case may be any deterioration in the condition of the furniture due to any ill-treatment by the tenant.

More significantly, it shall be assumed that the number of persons seeking to become tenants of similar dwelling-houses in the locality on the terms (other than those relating to rent) of the regulated tenancy is not substantially greater than the number of such dwelling-houses in the locality which are available for letting on such terms ... (see Baum (2007)).

The determination of a fair rent is governed by strict rules and regulations but, for our purposes, if the income approach is to be used to estimate market value, then the current fair rent and any future increase in the fair rent will be the income to be used in a capitalisation of such tenanted properties to assess the market value as an investment. It must also be noted that a fair rent can only be increased after two years and that there are complex rules relating to succession following the death of the original tenant (Schedule 1 of the RA 1977 as amended by Schedule 4 of the HA 1988).

However, there is still an active market for this type of investment, particularly where a portfolio of such properties is offered for sale. Those active in this market will be aware that a percentage of vacant possession value, between 25% and 50% in most cases, is used to assess market value when the income approach might only produce realistic market valuations by the use of very low capitalisation rates.

In a number of special cases, the tenancy will be held to be out-side the RA 1977 or HA 1988 and HA 1996 or, provided the correct notices are served at the commencement of the tenancy, the courts have a mandatory power to grant repossession. Examples of such tenancies are:

- holiday lettings
- lettings by educational establishments to students' lettings by absentee owners
- lettings of properties purchased for retirement.

In those cases where it is reasonable to assume that vacant possession can be obtained, the property should be valued by direct capital comparison. However, some consideration should be given to any time delays in obtaining possession.

## Tenancies with high RVs

The total number of tenanted houses and flats falling within this category represents a very small percentage of the whole, but it still comprises an important sector of the market, particularly in central London. To be unprotected and so outside of the RA 1977 regime, the RVs or rent must be higher than those set out above, under the heading 'Tenancies subject to the provisions of the Rent Act 1977'.

The valuation of such properties is by capitalisation of the current contracted rent with or without a reversion. The capitalisation rate must be market based and the rent will also be compared to other rents for similar properties.

## Resident landlords

The RA 1974 introduced a new special class of tenancy for lease after 14 August 1974 where the landlord resides in the same building. Such tenancies, whether of furnished or unfurnished premises, now fall under section 12 of the RA 1977.

The important feature of a resident landlord letting is that, although the tenant enjoys limited protection from eviction (RA 1977 Act, Protection from Eviction Act 1977 amended section 31 of the HA 1988), the landlord or, on the landlord's death, the landlord's personal representatives, can recover possession. If vacant possession can reasonably be expected then valuation by direct comparison with similar vacant possession properties will be the appropriate approach.

## Tenancies on long leases

In the case of many residential properties, the lease(s) are long leases and the tenants have valuable enfranchisement rights. This means that they can require the landlord to sell them the freehold interest or to grant a lease extension and, in the case of blocks of flats and similar multiple let long lease arrangements, to acquire freeholds collectively. This topic area is considered in detail in Baum (2007) and Shapiro et al (2009); it is more precisely an assessment of enfranchisement price in accordance with statutory rules and case precedent rather than an income approach.

### Question 1: Landlord and tenant issues

Your client owns the freehold interest in an office building. The building has a ground floor and four upper floors and the whole is let on full repairing and insuring terms for a term of 21 years without reviews. The lease has 3 years to run at £150,000 a year.

The building originally contained only a ground and two upper floors but the tenant agreed as a condition of the lease to add a third floor. This work was completed within 2 years of the grant at a cost of £150,000. Seven years ago a mansard roof was added at a further cost of £50,000. All these works were approved by the landlord.

There are 400 m$^2$ on each of the ground, first, second and third floors but the top floor has an area of only 250 m$^2$. Office space is letting at £200 per m$^2$. Where there is no lift, the rent on upper floors is £150 per m$^2$. The property is assessed for rating at RV £350,000. Freehold equivalent yields are 7%.

The tenants now wish to refit and equip the building with new carpets, lighting, computers, etc. but before proceeding wish to improve their security of tenure. They would prefer to buy the freehold but as a second best would like to surrender their present lease for a new 20-year full repairing and insuring lease with reviews every 5 years.

Advise the freeholder, (ignoring VAT):

a) on the rent that could be expected in 3 years' time;

b) on the price that should be asked for the freehold interest;

c) on the rent that should be asked for the proposed 20-year lease;

d) on the amount of compensation due to the tenant if possession was recovered.

# Chapter 10

## Development Opportunities

## Introduction

Development projects require the income approach to be incorporated in two valuation methodologies, the residual valuation and the development appraisal. Both require the valuer to assess income from a property not yet built, potentially on a site not yet purchased and therefore the valuer is not only making assumptions about market behaviour some way into the future; but is doing so in an environment of increasing uncertainty and risk.

Both approaches require the valuer to estimate the future value of the development with reference to income from rents and capital values on completion of the project and take away all of the costs of the project, including finance. Some professional guidelines require the rents and values to be based on current values (or at the valuation date), whereas in practice and in some situations explicit projections of costs and revenues are recommended.

When a developer is considering purchasing a site to develop a project, the valuation will be used to determine the price to be paid for the site. The valuer will estimate the project's value based upon what could be built under planning and other regulations and deduct all costs associated with the project, including a target profit set by the client.

When a developer is undertaking an appraisal where the land has been acquired or the price set, the same process as above will be applied. However, in this case, the land value will be included as a cost and not in the profit. The residual will now equal any profit (or loss, if negative) and will be used to test the viability of a scheme or alternative schemes.

So to summarise, the valuation of development projects is based upon two approaches using the same fundamentals:

| Value of Completed Development | less | Costs associated with the development | = | Surplus (Profit) or Deficit (Loss) |
|---|---|---|---|---|

- Where a valuer is tasked to determine what should be paid for a site, the surplus will be the land value and a target profit is included as a cost; this is called a *residual valuation*.

- Where a valuer is tasked to determine the profitability of a scheme, the surplus will be the developer's profit and land value is included as a cost; this is called a *development appraisal*.

The need for development appraisals or residual valuations occurs in a number of different contexts, for example:

- a bare or underdeveloped site where planning permission for development has been, or is likely to be, obtained
- an existing building for refurbishment or where planning permission has been, or is likely to be, obtained for a refurbishment and/or change of use
- in the case of an existing building where planning permission has been, or is likely to be, granted for its demolition and replacement.

In practice, the methodology adapts to different contexts, usually with increasing complexity and certainty as a scheme progresses. When used to calculate the price to pay for a development site, the input variables to the appraisal may be based on broad assumptions and generalised data, for example, building costs based on published indices and historic datasets. Later on, when the site has been purchased and construction is ongoing, detailed cost information on a monthly cash flow basis may be provided by a quantity surveyor and used in the appraisal to monitor project viability, thereby increasing its accuracy and complexity. The problems and criticisms of the initial appraisals are discussed later in this chapter.

Development costs are likely to be made on a monthly or even shorter cycle, and so a detailed cash flow is required to ensure an accurate picture of costs and revenues, especially on complex schemes which may have a long period of phased development.

The income approach requires the capitalisation of the potential future income of a scheme as part of a risk and return analysis, which is often undertaken at the site acquisition stage. The direct comparison method is preferred where possible and should be used as a 'reality check' when assessing site values. However, it is in this very market for property with development potential that the most difficulty in finding good comparable evidence is experienced.

## Incorporating a 'big picture' approach

Development valuations, especially on major projects, require the valuer to incorporate a significant and disparate amount of inputs, some direct, others more intangible and indirect, and the authors suggest that a 'big picture' methodology is incorporated to ensure that all aspects of the project and their direct or indirect impacts are considered. This is diagrammatically shown below as a 'jigsaw puzzle' where all elements of the development link together in terms of their impact upon the development viability analysis.

**Figure 10.1** The 'big picture approach' to development viability analysis.

Elements of the big picture approach include:

**Understanding the customer:** this is essential and requires a thorough analysis of the 'shape of demand', i.e. what the various subsectors of the markets identified require from a development. This may include finishes and appliances in residential property or the ability to create contemporary office layouts and space plans in office premises. Increasingly, the demand for sustainable development should be understood.

**National and regional agendas:** these include planning frameworks, sustainability targets, regeneration and other initiatives which may provide both costs and financial support for projects.

**Local politics, planning and competition:** these recognise that there will be local planning and other agendas. In some cases a change of political control in a local authority within the timeframe of a scheme may impose significant change. It is also essential to understand the potential competition for the anticipated development in terms of schemes both currently under construction and in the planning pipeline.

**Economics:** this should increasingly be at a global, national and local level of analysis as markets are intertwined and may impact upon development in many ways, including the availability and cost of finance. In a capital city occupational demand may be international and finance for major schemes may come from international investors.

**Finance:** the need to recognise the source and cost of finance. Increasingly this is not on a purely debt basis, and if equity is provided there may be complex arrangements for the equity provider to share in future profits. In major schemes a complex consortium of investors may be involved, each with their own profile of inputs and outputs from the scheme. Historically, in many cases development is 'forward funded' using long-term finance from an investor such as a pension fund or a consortium of private and/or corporate investors. Developers may also require short-term funding to finance the costs of construction before the project is sold to an investor.

**Infrastructure:** public transport in particular and the provision of local services are very important, and if not available immediately may impact upon the lettability or saleability of a scheme. Infrastructure may have to be provided as part of the planning requirement and therefore the cost and timing of appropriate infrastructure to support the development is essential.

**The market:** a very detailed analysis of market conditions will be required. This should include potential demand, vacancy rates and the absorption rate, that is how much property is being taken up over a period of time.

**The business case:** this has two dimensions; first, for investors it is the case for investing and will be dependent upon the risk and return profile of both the investor and the investment. Second, where the development, is for example, a leisure facility such as a golf course, the valuation must be based upon future projections of income from the business and will involve specialist consideration of projected incomes and costs.

**Marketing and letting:** this is becoming increasingly sophisticated and is used at different times of the development to raise awareness through to closing deals. This may include television advertising, promotional videos and DVDs, brochures, open days and launches.

**Law:** the legal position is an important constraint and may impact significantly upon the development process. For example, rights of way, easements and boundary issues may directly affect the scale, massing and layout of the design and restrict value. Existing tenants may also have legal rights and legal impositions may flow from planning permissions and funding agreements.

**Management:** it is important to consider the future asset management of a scheme. Creating a scheme with onerous management issues and complexity may impact upon an investor's view of a scheme as they will take over potential problems.

Finally, it is the **appraisal or valuation** that has to bring all of these elements together and quantify input variables which are heavily influenced by all of the other elements discussed above. All of the above sits in a context of rapid and constant change.

## Garbage in – garbage out

Development appraisals and especially residual valuations are notoriously unreliable and have been frequently criticised by the Lands Tribunal and in court judgments.

The adage 'garbage in – garbage out' is particularly pertinent to these appraisals as:

- *They involve an element of forecasting*, as the calculations are undertaken some time before the development is completed and lettings or sales are completed. In the case of complicated development projects completion may be some years into the future due to the need to obtain planning consents, complete infrastructure works and undertake building operations

■ *They are complex* and involve many variables which are highly sensitive to changes in both macroeconomic conditions and local micro-economic factors.

Where the residual method is used to assess the value of bare sites for rating or statutory purposes, it is often divorced from the realities of actual development. It has been heavily criticised by the Lands Tribunal, whose members denounced the method as 'far from a certain guide to values' (*Cuckmere Brick Co Ltd and Fawke* v *Mutual Finance Ltd* (1971) 218 EG 1571) and have suggested that 'once valuers are let loose upon residual valuations, however honest the valuers and however reasoned their arguments, they can prove almost anything' (*First Garden City Ltd* v *Letchworth Garden City Corporation* (1986) 200 EG 123).

Recently, the Lands Tribunal reconfirmed its distrust of the residual approach in a case involving the compulsory purchase of a development site. In *Ridgeland Properties Ltd (Claimant)* v *Bristol City Council (Acquiring Authority)* [2009] UKUT 102 (LC), the following extract from the ruling restates the Lands Tribunal's concerns with this methodology:

> Finally we would draw the parties' attention, as we did several times during the hearing, to the need to consider pragmatically and sensibly how much information a developer would expect and require in order to formulate an open market bid at the valuation date using the residual method of valuation. This Tribunal has repeatedly stressed its reluctance to use this valuation method. Its enforced use in this reference does not mean that its faults are any the less; it remains a valuation method of last resort which is inherently very sensitive to even small changes in the input variables.

This case highlighted a number of areas where different assumptions had made huge differences between the valuations of each party. For example, a difference of £4.5 m between the parties regarding the cost of superstructure works, due mainly to the difference in the experts' opinions about the cost of curtain wall cladding.

However, as will be explained later in this chapter, when the technique is used to appraise projects where land has been purchased for a known sum and a quantity surveyor has provided accurate cost information, the technique can become more robust.

# Professional frameworks and methodologies

A full consideration of the development appraisal process is outside the scope of this book. However, before considering the income approach in relation to development opportunities we feel it is wise to discuss preferred methodologies and professional frameworks to development appraisal, which help to reduce uncertainty and address some of its critics' concerns. The first is a *sifting* methodology which filters potential uses and is a precursor to the financial analysis of a development scheme for particular use or combination of uses, embodied in the North American Uniform Standards of Professional Appraisal Practice (USPAP) Regulations. The second is the Royal Institution of Chartered Surveyors' (RICS) approach, incorporated in Valuation Information Paper 12. Later in this chapter, we discuss in detail the issue of sensitivity to changes in variables and how techniques can be used to explore these sensitivities and make them explicit.

## The highest and best use methodology

A development appraisal can be required in a number of different circumstances, depending on these, the emphasis given to various aspects of the solution will vary. The most common development scenarios are where a developer owns or has an option on a site (which may be greenfield, brownfield, cleared or contain existing buildings) and we wish to determine the optimum use for it.

Conversely, we may have a development scheme in mind for which we know there is a proven market demand (e.g. a development of upmarket apartments or a mixed use scheme) but the developer needs to find the best location. Although differing in certain aspects, these are two sides of the same coin and both involve a good deal of similar work. Crucial to both sets of circumstances is the recognition of the need to find the optimum use for specific sites or locations. In North American appraisal practice this concept of 'optimum' or 'highest and best use' is the litmus test against which all proposals are judged. Essentially, North American Appraisal Regulations require the valuer or appraiser to advise the client of the most profitable concept which is feasible for a given site. That concept of feasibility is judged against three criteria:

- technical feasibility
- legal feasibility
- market feasibility.

In the United Kingdom we are not quite so explicit in our view, but the old adage that the success of a development rests on location, location and location, does express the reverse of the coin. The task is to tie these concepts together in a rational and convincing manner.

### The method

The highest and best use represents an explicit, logical and robust approach to refining the potential uses for a site. In theory any given site has a range of possible uses. The extent of this range is limited by various considerations such as site area, accessibility, slope, stability and surrounding uses. It may be quite possible, therefore, to build housing, shops, leisure facilities commercial or industrial, in fact any number or mix of uses on the site. It should be recognised that:

- some of the proposed uses will produce a higher rate of return than others

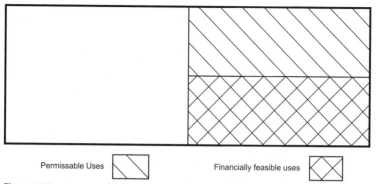

Permissable Uses    Financially feasible uses

**Figure 10.2**   How we refine the highest and best choices from the list of all possible uses

- not all represent feasible solutions in this location
- not all types of development will be granted planning permission.

Appraisers need, therefore, to refine their list of possible solutions to take these conditions into account.

Figure 10.2 below demonstrates the process. The outer bold rectangle illustrates all the possible uses of the site.

Starting from the total possible uses (the largest box) which can be produced by brainstorming, the uses can be refined first by identifying those possible uses which are permissible (the two right hand boxes). This process involves consideration of the development plan and perhaps discussion with the planning authority, as well as recognition of any other legal constraints on the site such as covenants, rights of way, etc. Under the US system this represents the legal feasibility evaluation stage. Next, we narrow our choice further by considering financial feasibility. At this point we would need to examine the macroeconomic climate and local market and sub-market indicators, taking into account interest rates, the property cycle, current and projected supply and demand, and local market conditions. The calculation would also involve consideration of development costs, including site acquisition, clearance, construction and/or refurbishment set against the development value of the completed scheme. Only those possible uses which fall within the bottom right hand box would therefore be worthy of detailed consideration. In this way appraisers and development consultants can demonstrate to their client that they have taken a rational approach and considered all the angles. It is of course what many practitioners/developers do unconsciously, but we would advocate an explicit approach using the above methodology.

## RICS Valuation Information Paper No 12: 'The Valuation of Development Land'

The RICS has produced a working paper on the valuation of development land and is also considering development viability planning methodologies. The paper aims to support the valuer and recognises that in valuing land for development, both a comparison and residual method can be used.

In the section entitled 'Establishing the Facts', the paper provides a useful checklist of issues to be considered, derived from the physical inspection and desk-based research to support the valuation. It includes ground conditions including potential contamination; capacity of infrastructure; evidence of rights of way, overhead power lines, water courses, etc; presence of archaeological features and many others. It also provides a thorough examination of existing planning matters to be considered from national to local and including special controls such as tree preservation orders, listing and environmental issues such as sites of special scientific interests.

The next section discusses the assessment of development potential and recognises that where the existing use or permissions is not optimum (the highest and best use), the valuer needs to form a view as to the form and extent of development that could be approved. The paper indicates four essential matters to be considered:

- The anticipated time period.
- The need to achieve a high efficiency ratio.
- Social and environmental issues that may affect the success of the project, including features that may be required to support the planning application.
- The extent to which the planning system is being used to help deliver climate change obligations.

This section also identifies the key components of the development programme, including the pre-construction period; the principal construction period and the post construction period and what activities are likely to take place in each. In the 'Analysing the Market' section it provides a useful indicative list of considerations:

- Owner occupier's current demands and preferences, for design, sustainability, internal layouts, etc.
- Investor's requirements in managing risk.
- Location issues.
- Access and public transport availability.
- Car parking.
- Local amenities attractive to potential tenants/purchasers such as cafes, banks, ATMs, etc.
- The scale of development and its divisibility into phases.
- The form of development.
- Market supply including actual or proposed developments.

In the 'Residual Method' section the guidance makes some useful comments about each of the inputs to the valuation. It is outside the scope of the book to list this in detail and we recommend readers access a copy of the paper. However, we have summarised some of the key points in the following paragraphs.

The gross development value is explicitly defined as the market value of the proposed development assessed on the special assumption that the development is complete as at the date of valuation in the market conditions prevailing at that date.

Under 'Development Costs', the paper indicates that it is necessary to allow for:

- obtaining planning permission and the cost and timing of any legally binding agreements associated with such permissions, for example the provision of social housing
- the potential need to provide environmental costs including sustainable urban drainage
- compliance with health and safety and site safety requirements and legislation
- securing the site and providing on site advertising.

When discussing building costs, the choice of procurement route is considered as this will influence the cash flow, and the duties and responsibilities of the contractors and the contingency arrangements.

An exhaustive list of potential fees and expenses is provided, including professional fees such as architects' and engineers' fees and expenses such as rent free incentives which impact upon the cash flow and the costs of raising the development finance.

In 'Assessing the Land Value', the report recommends that the sensitivity of the inputs be tested, which we discuss later in this chapter and to use the comparative method where possible as a 'reality check' as some residual valuations will produce highly theoretical results which deviate from market reality. The paper also highlights the fact that the residual approach may produce a negative value, which should be reported, although in some cases sites will sell in the market even where they may have a negative residual valuation.

## The residual method

In the valuation of sites and/or property with development potential, an adaptation of the income approach, known as the 'residual method', is the most widely used technique.

The conventional approach to a residual valuation is based upon the very simple concept introduced above; that the value of the completed project less all of the costs incurred in the project is equal to the residual value. The value of the completed project is usually known as the gross development value. This is the capital value of the finished development and is usually found by capitalisation of the projected income from the scheme. To confirm:

Gross development value less all costs and profit = site value

The costs of development include all of the costs of construction, including professional fees (architects', quantity surveyors' fees, etc), the cost of borrowing the capital to undertake the scheme, and the developer's profit. The surplus that remains after deducting all of these costs from the projected value is the residual value of the bare site or site and buildings to be redeveloped.

## Exploring the main inputs to a residual appraisal

It has already been suggested that one of the difficulties in applying the residual method revolves around the number and variability of the inputs, these are introduced below.

### Completed development value

This is the market value of the completed development which would be arrived at by the comparison or investment method as appropriate, depending upon the type of property.

In the case of commercial developments, it is assumed that on completion, the developer will sell the development to an investor. This will incur agents' fees, advertising and legal fees which may amount to approximately 3% of the value of the completed development. In addition it will be easier to sell if the building is already occupied and so letting agent's fees, which might account for some 10% of the first year's rent, also need to be taken into account.

### Rent and yield

In the case of most commercial/industrial buildings, the completed development value will be calculated using the investment method of valuation. This will require an estimation of the rent of the development and the likely investment yield. Both will be based on current market evidence. The rent is normally based on the lettable area of the completed development. This will usually be the net internal area. It should be noted that in a residual appraisal it is the current rent which is used, notwithstanding the fact that rental levels may have changed by the time the development is completed. It is regarded by some as extremely dangerous to base an appraisal on inflated future rents, as there can be no guarantee that these will be achieved. Others would argue that a full explicit cash flow approach requires the modelling of rents based upon projections both up and down over the anticipated period of construction and letting. The possibility of rental increase/ decrease over the period of the development is something which can be reflected in the risk or profit allowance.

### Incidental costs

It is assumed that the developer will dispose of the completed development and will thus incur incidental disposal costs which will include agent and legal fees amounting to around 2.5%. The developer will also incur costs in acquiring the site, including land value stamp duty currently at 1 to 4%, legal fees, typically between 0.25 and 0.5% and agents' fees typically of 1 to 2%.

### Costs

The main costs to be deducted from the completed development value will include pre- and post-development costs.

**Pre-development costs.** The pre-development costs are those occurring before the commencement of construction, and typically include:

- demolition
- provision of infrastructure such as roads, electricity and gas supplies
- planning gain including cash contributions or construction works, for example a road junction
- compensation to tenants to remove them if they have security of tenure
- site investigation
- ground investigation
- land surveys
- planning fees
- planning consultant fees
- legal costs for appeals
- building regulations, scale fee based on final building cost
- funding fees.

**Construction costs.** Construction costs in initial appraisals or residual valuations are typically assessed on a price per m² basis. Sources of cost information include *Spon's Architects' and Builders' Price Book* which is published annually, and the RICS Building Cost Information Service. In practice advanced appraisals will be based upon detailed on building costs identified by a quantity surveyor.

**Fees.** The professional fees for architects, QS, Engineers, etc. need to be added to the construction costs. It is typical to add 10 to 15% to the construction costs, but this will be dependent upon the nature and scale of the development. In more complex schemes it may be necessary to make additional allowance for project manager's fees, structural and engineering consultants and specialists, such as mechanical engineers, if the building is air conditioned or, for example, heated by sustainable technology such as ground source heat pumps.

**Finance.** The developer will need to take out short-term loans to fund the development. Even if the developer is able to fund the development from capital, this will incur opportunity costs. Funding needs to cover the length of the development period, which must reflect the pre-contract period at the outset and any letting or void period once the development is completed. It is only on successful completion and sale that the developer will be able to recoup costs.

The basic residual usually adopts a 'rule of thumb' approach to the calculation of finance charges by either taking half the development period or half the building costs. This is usually adopted to reflect the average cost. Clearly, the developer will not borrow the whole of the cost of the development at the beginning of the development period, as payments to contractors will be staged as the project progresses.

The costs of short-term finance vary dramatically with economic conditions and the strength of the borrower in terms of risk. They have historically been between 1 and 6% above the prevailing base rate. The actual rate will depend upon the status and track record of the developer, as well as the type, quality and location of the development.

The developer will also need to allow for the cost of interest payments on the site acquisition price. Where a site is being purchased at the outset, the cost of this must be funded over the whole period of development, from purchase through to disposal. In valuations, finance is typically assumed to be 100% debt based, in reality it is more likely to be a combination of equity and debt.

***Marketing/letting promotion.*** It is necessary to allow for both the letting and sale of the completed development. This will include:

- advertising
- particulars and brochures
- site boards
- mail shots
- launch ceremonies
- show suites
- public relations
- letting incentives.

***Miscellaneous costs.*** Other costs which need to be reflected may include:

- party wall agreements
- rights of light agreements
- void periods – maintenance and insurance
- Rent free periods
- empty rates.

Some costs are difficult to identify or estimate accurately and these are sometimes reflected in a contingency of around 3 to 5%. It is sometimes argued that these elements are included in the profit/risk element. Thus, in the case of a development with major cost uncertainty the developer will seek to make a higher allowance for profit.

### Developer's profit or risk

If no allowance was made for profit, the developer would be undertaking the project for nothing. This allowance is normally related either to the cost of the development or the completed development value; typically in the United Kingdom this might be 15 to 25% cost, or 10 to 20% value. In large schemes and in other countries these rates may be less.

If the market is highly competitive, developers may be forced to trim profit margins, however if the risks are high, for example if the time taken to let or sell the development is extended in a weak market, then the profit element would need to be increased.

**Example 10.1**

A plot of land has planning permission for the erection of 7,000 m² (gross) of office space on five floors.

The development will be completed in two years from now and rents are expected to average £120 per m² (net).

Building costs are expected to average £400 per m², excluding fees.

Prepare a valuation to advise a prospective developer of the maximum bid to be made for the site.

(Comparable evidence of prices obtained for similar sites is not available.)

Gross development value:

| | | |
|---|---|---|
| 7,000 m² (gross) × 85%* | = (say) 6,000 m² (net) | |
| 6,000 m² at £120 per m² pa | = £720,000 pa | |
| × capitalisation factor @ 7% | 14.29 | |
| | | £10,285,714 |

*Reduction for non-usable space of 15%.

Gross development value (GDV): c/o £10,285,714

Less costs:

(a) Building costs:
    7,000 m² at £400 per m²  £2,800,000

(b) Architects' and quantity surveyors'
    fees at 12.5% of building costs  £350,000

    Total building costs:  £3,150,000

(c) Finance costs:
    15% pa for 2 years on 50% of total
    building costs (£1,575,000 × $(1.15)^2$)
    − £1,575,000 =  £507,937

(d) Legal fees, estate agents' fees
    and advertising upon disposal:
    at 2% GDV = £205,714

(e) Promotion say:  £50,000

(f) Developer's profit: @ 15% GDV  £1,542,857
    Total costs:  £5,456,508

Residual: (GDV − costs)  £4,829,206

Let site value = $x$

Legal and valuation fees on site purchase at 3% = $0.03x$

Total accumulated debt after 2 years at 15% pa = $1.03x (1.15)^2$

| | | |
|---|---|---|
| $1.03x (1.15)^2$ | = | £4,829,207 |
| $1.362x$ | = | £4,829,207 |
| Site value | = | £3,545,673 |

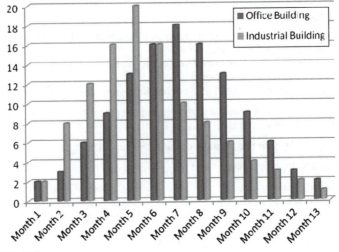

**Figure 10.3**

As stated earlier, the use of the residual method varies according to the stage of the development process. The above calculation is much generalised in its assumptions and the technique is often criticised as a method of determining land value. Example 10.1 can be used to illustrate the dangers of the generalised assumptions built into the simple residual valuation.

For example, the finance for development is calculated on only 50% of the total construction costs to reflect the phased drawing down of borrowing over the development period. This assumption will hold true if the payments to contractors follow an equal distribution pattern similar to that shown in Figure 10.3 for the traditional office building. However, if the building involved a portal steel frame which brings the development costs forward, the expenditure pattern may change to that shown in Figure 10.3 for an industrial building. This destroys the 50% assumption and an expenditure of this pattern should assume funding on around 75% of the total building costs.

The need for accurate, cash flow based projections of building costs are therefore essential for accurate residual valuations. The RICS Valuation Information Paper No 12 recognises this issue and states that an 'S-curve' may be applied, the S-curve is the shape of the above Figure when the amounts are made cumulative. Each development will have its own 'signature' S-curve depending upon its use, construction and procurement options and these can be evaluated by a quantity surveyor or cost consultant.

## Sensitivity analysis

In addition to being subject to generalised assumptions, the final value is very sensitive to small variations in key variables, particularly the yield, rental values and building costs.

To illustrate this problem note the effects on the residual value calculation in Example 10.1 with the changes to the key variables as shown in Figure 10.4:

| | |
|---|---|
| Original site value | £3,545,673 |
| Site value when yield only increased by 0.5% | £3,120,825 |
| Site value when rents only increased by 10% | £4,171,945 |
| Site value when building costs only increased by 10% | £3,276,680 |

**Figure 10.4**   Illustration of the sensitivity of the residual site value calculated in Example 10.1 to changes in key variables

The use of 'STAR' diagrams to analyse relative sensitivity is also an increasingly common practice and one advocated by the authors. Following financial modelling principles, it maps onto a single line chart the variation in returns against changes in a number of variables. The chart is therefore a quick and easy way to read what a project is most sensitive to through the steepness of the lines, which indicates the change in profit against the change in variables. This methodology requires *all* changes in variables to be measured in percentage terms using the same percentage step for each variable. Some computer packages, such as ProDev, allow you to do this and create a table of values or returns against changes in variables.

Using the information in Example 10.1 again we can create a table in Figure 10.5 of site values using changes in the main underlying variables:

| | Building Costs | Site Value | Office Rents | Site Value | Yield | Site Value |
|---|---|---|---|---|---|---|
| -20% | 320 | 4082290 | 96 | 2291759 | 5.6% | 5112039 |
| -15% | 340 | 3948021 | 102 | 2605123 | 6.0% | 4589765 |
| -10% | 360 | 3813753 | 108 | 2918488 | 6.3% | 4241582 |
| -5% | 380 | 3679485 | 114 | 3231852 | 6.7% | 3825841 |
| **Original Figures** | **400** | **3545673** | **120** | **3545673** | **7.0%** | **3545673** |
| +5% | 420 | 3410948 | 126 | 3858581 | 7.4% | 3206444 |
| +10% | 440 | 3276680 | 132 | 4171945 | 7.7% | 2975463 |
| +15% | 460 | 3142412 | 138 | 4485310 | 8.1% | 2694103 |
| +20% | 480 | 3008143 | 144 | 4798674 | 8.4% | 2500669 |

**Figure 10.5**   A table of sensitivity values

These can then be turned into a 'star chart', as shown in Figure 10.6, to demonstrate the relative sensitivity of the site value to changes in the underlying variables.

The chart shows that rents and yields are the most sensitive variables as the line is steeper, indicating that for the same change in the variable there is a greater change in the residual value than for building costs.

A further problem facing valuers has been the difficulty of handling an increasing number of variables such as VAT, fees, etc in the valuation exercise when carried out manually. Currently, many

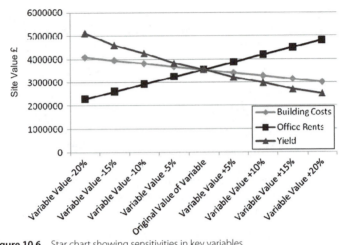

**Figure 10.6** Star chart showing sensitivities in key variables

development surveyors use spreadsheets or computer software packages which do not remove the basic criticism, but their flexibility and speed of operation enable valuers to check each appraisal for sensitivity and avoid generalisations in the assumptions to achieve greater confidence in their opinions of value. (An example of this approach is provided in Appendix C.)

Example 10.1 has been reappraised using a package prepared and marketed by Argus Software, which is one of several programs now available to development valuers. The report produced by the system is set out in Appendix C, together with notes on the assumptions and methodologies used.

The traditional residual approach can therefore be significantly improved with the application of computers through sensitivity analysis and more complex treatment of the underlying assumptions. However, although packages such as Argus allow for complex phasing of developments by the linking of parts and areas of schemes together, they still rely heavily on assumptions made about the distribution of payments and receipts.

## The need for adoption of the cash flow approach

Cash flow modelling is now a standard feature of most development appraisal systems and its use is encouraged to counter the difficulties and criticisms described above.

These are especially useful when dealing with large, phased developments such as business parks or major residential schemes. The cash flow approach also allows the valuer to take into account the financial status of the client and to provide a more accurate and realistic financial appraisal. For example, joint ventures or schemes involving equity finance may require differential interest rates and complex flows of income which cannot be accommodated in a traditional appraisal.

As the valuation profession is being encouraged to be more conscious of the needs of its clients, the cash flow approach should be more widely adopted to ensure that the financial situation of the client is accurately modelled within the appraisal.

## The cash flow approach explained

Testing the traditional residual results against cash flow models suggests that the greater variability occurs with phased developments such as low-rise residential schemes, where cash in-flows coincide with cash out-flows. The variation in result is less pronounced with the relatively simple one-off development, especially those with a project term of 12 months or less. In the absence of alternative development schemes, it is almost certain that the cost of borrowing money will be at least as high as the rate of interest that the developer could earn on excess funds. It is therefore reasonable to conclude that the developer should pay off debts as soon as income is produced by the development.

The developer will pay considerable attention to the likely cash flow and the likely maximum borrowing requirement over the building period. A cash flow table will provide full information concerning these two points, and will allow for the accurate estimation of finance charges. In very complex multi-use schemes, the development surveyor and quantity surveyors may be able to use cash flow analysis to reschedule the detailed timing of development to improve the overall return on capital.

Systems such as Argus allow the user to construct a monthly cash flow from the original residual valuation which can then be edited to reflect more accurate data provided by quantity surveyors, or to explore the effects of adjusting the cash flow. The program provides the user with a full discounted cash flow (DCF) analysis of the project and can also be converted into a spreadsheet files.

The example below illustrates the cash flow approach using a simple manual calculation.

---

**Example 10.2**

A site has the benefit of planning permission for the erection of a block of 20 apartments, totalling 1,100 m² of space.

The value of each completed apartment is estimated to be £100,000, using comparable transactions, but flat prices are expected to continue to rise at 5% pa.

Building costs are estimated at £800 per m² and are expected to rise at 0.5% per month. It is expected that five apartments will be sold in each of months 9–12 and that building costs will be evenly distributed over a 12-month building period.

Architects' and quantity surveyors' fees will be payable in two instalments in months 6 and 12 at 10% of the building costs.

Agents' and solicitors' fees will be charged on sale at 1% of the sale price of each apartment, and the developer will require a profit of 15% of the sale price.

Advise the developer as to the maximum bid for the site.

1 Apartment prices are increasing at 5% pa or 0.4% per month.

In month 9,
5 flats will sell for £103,660 each   = £518,300

In month 10,
5 flats will sell for £104,070 each   = £520,350

In month 11,
5 flats will sell for £104,500 each   = £522,500

In month 12,
5 flats will sell for £104,910 each   = £524,550

2 Building costs are estimated at £800 per m², spread evenly over 12 months, rising at 0.5% per month.

Total building costs for the entire block of £880,000 divided by 12 gives a cost of £73,330 in month 1 which will increase by 0.5% in month 2, etc.

This results in a building cost flow of:
    **month 1: £73,330**
    **month 2: £73,700**
    **month 3: £74,060**
    **month 4: £74,440**
    **month 5: £74,810**

in month 6 fees are payable on the costs to date
at a rate of 10%
(10% of £445,490 = £44,549) £75,180 + £44,549 = **month 6: £119,729**
    **month 7: £75,560**
    **month 8: £75,940**

in month 9 the first of the sales of 5 flats is achieved adding fees at 1% and profit at 15% of the sale price to the costs
(1% of £518,300 = £5,183 and
15% of £518,300 = £77,745)

£76,320 + £5,183 + £77,745     **month 9: £159,248**
in month 10: £76,700 + £5,204 + £78,053
(calculated as above but on
month 10 sale prices)     **month 10: £159957**

in month 11: £77,080 + £5,225 + £78,375     **month 11: £160680**
(see above)

in month 12: £77,470 + £5,246 + £78,683     **month 12: £161399**
    + £45,900 (Arch/QS fees @
    10% costs months 7–12)

From this information a cash flow table can be constructed incorporating the effects of interest charges at 1% per month.

| Month | Benefits(£) | Costs (£) Net (£) | Capital (£) | Interest (£) | Outstanding |
|---|---|---|---|---|---|
| 1 | – | 73,330 | –73,330 | –73,333 | 733 |
| 2 | – | 73,700 | –73,700 | –147,760[1] | 1,480 |
| 3 | – | 74,060 | –74,060 | –223,301 | 2,233 |
| 4 | – | 74,440 | –74,440 | –299,974 | 3,000 |
| 5 | – | 74,810 | –74,810 | –377,784 | 3,778 |
| 6 | – | 119,729 | –119,729 | –501,291 | 5,013 |
| 7 | – | 75,560 | –75,560 | –581,863 | 5,819 |
| 8 | – | 75,940 | –75,940 | –663,622 | 6,636 |
| 9 | 518,300 | 159,248 | +359,052 | –311,206 | 3,112 |
| 10 | 520,350 | 159,957 | +360,393 | +46,075[2] | +461 |
| 11 | 522,500 | 160,680 | +361,820 | +408,355 | +4,084 |
| 12 | 524,450 | 207,299 | +317,251 | +729,690[3] | |

Residual          $= £729,690$

Let site value      $= x$

Let fees on site purchase    $= 3\% = 0.03x$

Then

| | |
|---|---|
| $1.03x +$ interest for 12 months at 1% per month | $= £729,690$ |
| $1.03x(1.01)12$ | $= £729,690$ |
| $1.03x(1.1268)$ | $= £729,690$ |
| $1.1606x$ | $= £729,690$ |
| $x$ | $= £628,718$ |
| Site value | $= £628,718$ |

Notes:

1. Already outstanding from month 1 is a debt of £73,330, and an interest charge of £733. Added to these is a new loan of £73,700, the total being £147,763 on which the interest charge for month 2 will be levied.
2. This figure if positive, as the total debt of £311,206 plus £3,112 interest is more than repaid by the receipt in month 10 of £360,393. Hence a surplus of £46,075 remains. This can be invested to earn interest at 1% per month.
3. This final surplus remains after paying all costs except the cost of the site itself, and fees and finance on its purchase. Calculated as before, the developer's maximum site bid is £628,718.

## Example 10.3

In Example 10.2 the cash flow has been undertaken on a terminal value basis. In this example the same development is valued using a present value approach.

| Month | Benefits(£) | Costs (£) | Net (£) | PV£1 at 1% | PV Outstanding |
|---|---|---|---|---|---|
| 1 | – | 73,330 | –73,330 | 0.99001 | –72,604 |
| 2 | – | 73,700 | –73,700 | 0.98030 | –72,248 |
| 3 | – | 74,060 | –74,060 | 0.97060 | –71,882 |
| 4 | – | 74,440 | –74,440 | 0.96098 | –71,535 |

| 5 | | – | 74,810 | –74,810 | 0.95147 | –71,179 |
|---|---|---|---|---|---|---|
| 6 | | – | 119,729 | 119,729 | 0.94205 | –112,270 |
| 7 | | – | 75,560 | –75,560 | 0.93272 | –70,476 |
| 8 | | – | 75,940 | –75,940 | 0.92348 | –70,129 |
| 9 | 518,300 | | 159,248 | +359,052 | 0.91434 | +328,296 |
| 10 | 520,350 | | 159,957 | +360,393 | 0.90529 | +326,259 |
| 11 | 522,500 | | 160,680 | +361,820 | 0.89632 | +324,308 |
| 12 | 524,450 | | 207,299 | +317,251 | 0.88745 | +281,544 |

NPV = £647,563

| | | |
|---|---|---|
| Let site value | = | $x$ |
| Let fees on site purchase | = | 3% = 0.03$x$. |
| Then | | |
| 1.03$x$ | = | £647,563 |
| $X$ | = | £628,702 (variation from example 11.2 due to rounding). |
| Site value | = | £628,702 |

An important prerequisite for a cash flow approach is an accurate scheduling of construction activities.

This uses network analysis, or critical path analysis whereby critical sequential events become the critical path and other non-sequential events can be scheduled to run in parallel; in turn some of the parallel costs may also be sequential.

The end result is the production of a project network with a start and finish date with all the intermediate critical dates. The actual costs can then be estimated and transferred to a cash flow. The two together will then become part of the project manager's management tools for monitoring the development programme. Such an approach, if applied to Example 10.1, would highlight the need to reduce the construction period in order to reduce the high finance costs. The need to reduce finance costs in periods of high interest rates has played its part in developing innovative 'fast track' construction techniques.

Cash flows and residuals can be calculated on a current cost basis, or on a future cost basis building in variations in rents, as well as variations in labour and materials. None of the computerised sophistications can, however, overcome the problem that the acceptability of residual and cash flow development appraisal methods rests not with their rationale, which is irrefutable, but with the quality of the evidence used by the appraisal team to estimate costs and benefits.

## Viability studies

Earlier in this chapter it was stated that the residual method of valuation is based upon a simple equation:

Gross development value – development costs (including profit)
= residual value

The aim of the residual valuation is to find the unknown in this equation; the residual value. Often, however, a prospective developer will be aware of the likely site cost and consequently of the cost of fees and finance, either because the vendor has stated an asking price, or because negotiations have revealed the minimum figure which the vendor will accept. This is particularly likely in the case of an existing building which is to be improved or replaced.

In such a case the single unknown in the equation can now become the developer's profit. The aim of a viability study is to assess the profit likely to be made from a development scheme, given the cost or asking price of the subject site.

Computer packages, for example Circle, allow the user to select from a drop down menu whether the system is to calculate residual land value or developer's profit. The input screens and processing are then adjusted according to the user's selection.

The cash flow approach is advocated by most software manufacturers, if not valuers, because of its greater accuracy and ability to model the payment of contractors, architects and finance charges much more closely to the actual payment profile than the smoothed assumptions often contained in a traditional appraisal.

In addition, a detailed cash flow approach allows the modelling of complex financial arrangements that may often underpin major property developments around the world. For example, equity sharing and joint ventures where the interest rates may differ for each project partner, depending upon their risk profile.

The cash flow table is particularly adaptable as a viability study and the authors recommend that valuers using appraisal software always utilise a cash flow approach, and only use the initial appraisal as a way of inputting data and obtaining a summary picture of the development analysis.

### Example 10.4

Using the facts of Example 10.2, advise the developer of the potential profit if the total cost of the site, including fees, is £500,000.

| Month | Benefits(£) | Costs (£) | Net (£) | Capital (£) | Interest (£) outstanding |
|---|---|---|---|---|---|
| 1 | – | 573,330 | –573,330 | –573,333 | 5,733 |
| 2 | – | 73,700 | –73,700 | –652,763 | 6,528 |
| 3 | – | 74,060 | –74,060 | –733,351 | 7,334 |
| 4 | – | 74,440 | –74,440 | –815,124 | 8,151 |
| 5 | – | 74,810 | –74,810 | –898,086 | 8,981 |
| 6 | – | 119,729 | –119,729 | –1,026,796 | 10,013 |
| 7 | – | 75,560 | –75,560 | –1,112,623 | 11,126 |
| 8 | – | 75,940 | –75,940 | –1,199,690 | 11,997 |
| 9 | 518,300 | 81,503 | +436,797 | –774,890 | 7,749 |
| 10 | 520,350 | 81,904 | +438,446 | –344,193 | 3,442 |
| 11 | 522,500 | 82,305 | +440,195 | +92,561 | +926 |
| 12 | 524,450 | 128,616 | +395,934 | +489,420 | |

£489,420 represents the developer's profit

This can be expressed as     $\dfrac{£489,420}{£2,085,700}$     =     23.47% of GDV

or as     $\dfrac{£489,420}{£1,515,897}$     =     32.29% of total costs

The residual method has become a straightforward DCF exercise. Nevertheless, there will still be areas of variability, and the criticism that the result is sensitive to changes in inputs remains. The next stage is therefore to incorporate probability measures in the analysis, such as Monte Carlo Risk Simulations as advocated in the RICS Valuation Information Paper No 12 (see Chapter 13).

Rents, costs and interest charges may be weighted according to the valuer's expectation of possible changes, such an approach being eminently suited to computer analysis.

### Question 1: Residual site value

Assess the residual site value of a parcel of land suitable for 5,000 m² of office space.

The scheme will take 12 months to complete. Rents are £200 per m² (net lettable area).

Construction costs are £400 per m².

The scheme would sell on a 7% basis and finance is available at 14%.

Promotion costs are budgeted at £40,000.

Letting fees are agreed at 15% of the first year's rent.

Purchases costs are 1.75%,

Site purchase costs 1.5%, stamp duty 1%.

Agents' fees are 1% of GDV.

Developer's profit required is 15% of GDV.

## Summary

- Residual valuations or development appraisals are used to assess the value of land and buildings with latent development potential.
- The method is based upon established economic principles and the method itself is logical.
- However, the accuracy of the residual method relies on the valuer's ability to estimate all of the variables within the calculation. The costs and benefits must be carefully estimated with respect to the development market, using hard evidence where possible to support the rents.
- Capitalisation rates and building costs adopted.
- Development appraisal is an ideal application for information technology, allowing the valuer to assess the sensitivity of the valuation to changes in the underlying variables and to apply simulation techniques to model the probability of outcomes.

- The final opinion of development value should, wherever possible, be checked against comparable market evidence. However, as each development opportunity is unique in terms of its highest and best use, market comparison will only provide the valuer with a basic 'yardstick' with which to cross reference the appraisal.

# Spreadsheet user

## Using Excel to build an appraisal

### Value

We will now explore a simple development appraisal using a cash flow approach by building an Excel spreadsheet to undertake the development appraisal.

### Appraisal data

Construction of 2000 m² (lettable area) offices.
Estimated rental value per m² on completion £1000 paid in advance.
Estimated time from completion of construction 12 months to full occupation.
Estimated take up rate: 500 m²/quarter.
Construction complete 12 months time.
Construction commences immediately.

Using this data we can begin to construct a development appraisal spreadsheet.

**(i)** Enter quarterly time periods across the spreadsheet:

| 0 | 1 | 2 | 3 | 4 | 5 | 6 | 7 | 8 |
|---|---|---|---|---|---|---|---|---|

**(ii)** Enter the rental income commencing in quarter 4 (remember that in Excel payments are treated as occurring at the end of the period when using net present value and internal rate of return functions).

**(iii)** Assuming that an investor has made an agreement to purchase the development when it is fully let at an agreed all risks yield of 8%.

We have now completed the (positive) value inputs into the development and your spreadsheet should be similar to the following screenshot:

Note: The sale proceeds have been calculated by applying the all risks yield to the total income when the development is fully let, ie:

$$2000 \text{ m}^2 \text{ at } \pounds 1000/\text{m}^2 = \pounds 2,000,000$$

$$\times \text{ Capitalisation Factor } \left(\frac{1}{i}\right) = \left(\frac{1}{0.08}\right) = 12.5$$

$$= \pounds 25,000,000$$

We have now considered the value side of the residual equation by considering the income flow during the development period. Now we need to consider the costs of development.

### Estimating the costs of development

The costs of development have been discussed in detail above and examples of costs during the pre-construction, principal or main construction period and the post construction period are given under both the RICS Valuation Information Paper No 12 and the section entitled 'The residual method' in this chapter.

### Building the costs into our Excel spreadsheet

Returning to our spreadsheet, we must now enter the following costs into our simple appraisal:

*Pre-development costs (all occurring in the first month).*

**1.** A planning fee of £10,000.
**2.** A building permit fee of £5,000.
**3.** A site investigation of £20,000.

*Construction costs.*

**1.** Construction costs:

£2,000,000 end of quarter 0
£6,000,000 end of quarter 1
£6,000,000 end of quarter 2
£2,000,000 end of quarter 3.

**2.** Contingency:

5% of construction costs.

**3.** Professional fees:

10% of construction costs (excluding contingency).

*Finance.* Finance is available at an interest rate (cost of capital) of 2.5% per quarter.

*Marketing.*

Agents letting fees: 7.5% of initial rent.

*Promotion.*

£ 50,000 quarter 0
£ 20,000 quarter 1
£ 10,000 quarter 2
£ 10,000 quarter 3
£ 10,000 quarter 4
£ 20,000 quarter 5

### Site costs (all payable in quarter 0).

**1.** The site costs of £4,500,000.

**2.** Agents and solicitors fees amount to £100,000.

**3.** Property tax of £50,000.

When you have entered the data into the spreadsheet it should look something like this:

| | B | C | D | E | F | G | H | I | J | K | L |
|---|---|---|---|---|---|---|---|---|---|---|---|
| 5 | Item | | | | | | | | | | |
| 6 | Rent | | | | | 500000 | 1000000 | 1500000 | 2000000 | | |
| 7 | Sale Proceeds | | | | | | | | | 25000000 | |
| 9 | Costs: | | | | | | | | | | |
| 11 | Planning Fee | -10000 | | | | | | | | | |
| 12 | Building Fee | -5000 | | | | | | | | | |
| 13 | Site Investigation | -20000 | | | | | | | | | |
| 14 | Construction | -2000000 | -6000000 | -6000000 | -2000000 | | | | | | |
| 15 | Contingency | -100000 | -300000 | -300000 | -100000 | | | | | | |
| 16 | Prof. Fees | -200000 | -600000 | -600000 | -200000 | | | | | | |
| 17 | Letting Fee | | | | | -37500 | -37500 | -37500 | -37500 | | |
| 18 | Promotion | -50000 | -20000 | -10000 | -10000 | -10000 | -20000 | | | | |
| 19 | Land Cost | -4500000 | | | | | | | | | |
| 20 | Fees | -100000 | | | | | | | | | |
| 21 | Tax | -50000 | | | | | | | | | |

Now we have to calculate the cash flow.

Use the Excel Sum function to add the figures in each column.

In our spreadsheet in cell C23 type (or use Autosum) = Sum (C6:C21).

This can then be copied across to all the rows for quarter 0 to quarter 8 using the copy and paste buttons/keys or the auto-drag option.

You now have the undiscounted cash flow.

### Calculating the net present value

Next:

Enter the formula for the present value (PV) into the cell below the first cash flow figure (in our spreadsheet C20).

$$= \frac{1}{(1+i)^n}$$

Remember you will need appropriate brackets and that we are using quarterly periods and a quarterly interest rate.

The contents of our cell C24 are:

$$(1/((1+0.025)\char`^C3))\,{}^*C23$$

This formula can then be copied across all the quarters.

You can then use the sum function to add all of the quarterly discounted cash flow figures to give you the net present value (NPV).

Alternatively, you can use the Excel NPV function [= NPV (discount rate, range of cash flow)] by typing in: = NPV (0.025, C23:K23)

If you use the above you will note that it does not give the same figure as calculating it using the PV formula and summing the discounted cash flows. This is because the function assumes the investment begins one period before the date of the first value and that the initial payment is made at the end of the first time period. In fact in our appraisal, as in most property cash flows, the site is purchased at the start of the first period. Therefore, the NPV function must be modified in our case to:

$$= NPV\ (0.025,\ D23 : K23) + C23$$

This function should now give you the £2,165,410 expected result.

You are advised to print out, read and file in this section the Excel help sheet on the NPV function.

The NPV represents the developer's profit and can be converted to a return on capital costs or value.

### Calculating the internal rate of return

The internal rate of return (IRR) can now be calculated using the Excel IRR function:

$$= IRR\ (range\ of\ cash\ flow, trial\ rate)$$

In our spreadsheet we have entered into cell C22 the formula:

$$= IRR(C23 : K23, 0.05)$$

$$= 3.960\ \%$$

Remember, a trial rate is needed (we have used 0.05) to initialise the process of iteration.

You are advised to print out, read and file in this section the Excel help sheet on the IRR function.

When you have calculated the NPV and IRR, your spreadsheet should look something like this:

| | A | B | C | D | E | F | G | H | I | J | K | L |
|---|---|---|---|---|---|---|---|---|---|---|---|---|
| 5 | | Item | | | | | | | | | | |
| 6 | | Rent | | | | | 500000 | 1000000 | 1500000 | 2000000 | | |
| 7 | | Sale Proceeds | | | | | | | | | 25000000 | |
| 9 | | Costs: | | | | | | | | | | |
| 11 | | Planning Fee | -10000 | | | | | | | | | |
| 12 | | Building Fee | -5000 | | | | | | | | | |
| 13 | | Site Investigation | -20000 | | | | | | | | | |
| 14 | | Construction | -2000000 | -6000000 | -6000000 | -2000000 | | | | | | |
| 15 | | Contingency | -100000 | -300000 | -300000 | -100000 | | | | | | |
| 16 | | Prof. Fees | -200000 | -600000 | -600000 | -200000 | | | | | | |
| 17 | | Letting Fee | | | | | -37500 | -37500 | -37500 | -37500 | | |
| 18 | | Promotion | -50000 | -20000 | -10000 | -10000 | -10000 | -20000 | | | | |
| 19 | | Land Cost | -4500000 | | | | | | | | | |
| 20 | | Fees | -100000 | | | | | | | | | |
| 21 | | Tax | -50000 | | | | | | | | | |
| 23 | | CASH FLOW | -7035000 | -6920000 | -6910000 | -2310000 | 452500 | 942500 | 1462500 | 1962500 | 25000000 | |
| 24 | | P.V@ at 2.5% | -7035000 | -6751220 | -6577037 | -2145065 | 409943 | 833033 | 1261109 | 1650983 | 20518664 | |
| 25 | | N.P.V = | 2165410 | | | | | | | | | |
| 26 | | I.R.R = | 3.960% | | | Effective Annual Rate = | | | 16.81% | | | |

Remember, this is a quarterly IRR and to convert to an effective annual rate you must use the formula:

$$(1 + i)^{nth} - 1 \times 100$$

where n is the number of periods per annum.

Therefore, in our example:

$$(1 + 0.03960)^4 - 1 \times 100 \ (4 = \text{quarterly periods})$$

$$= 16.806\%$$

In our spreadsheet we have entered into cell I26 the following formula:

$$((((1 + (C26))^\wedge 4) - 1)$$

This cell has been set to a percentage format; if your cell I26 is not in a percentage format your formula must be adjusted by adding *100 to the end of the formula.

# Chapter 11

## The Profits Method of Valuation

## Rationale behind the use of the profits method of valuation

With most types of commercial property it is possible to arrive at a rental value by considering evidence of similar properties on a rent per square metre (foot) basis, interpreting that evidence and then applying a rent per square metre (foot) to the subject property.

This is not possible with some classes of property as the rental value will be driven by the trading potential rather than purely by the location, size and specification of the demise, compared to other properties of the same class.

Properties which are valued with regard to trading potential include licensed premises such as public houses, bars, clubs, restaurants and hotels. They also include theatres, cinemas, care homes, caravan parks and petrol filling stations.

The Royal Institution of Chartered Surveyors (RICS) Valuation Standards, compliant with International Valuation Standards, enshrine the methodology. A new RICS Guidance Note (GN2) can be found in the RICS Valuation Standards 2011. It considers the *Valuation of individual trade related property* including market rent. It is compulsory reading for valuers and should be read by students.

The RICS has also published two editions of a Valuation Information Paper No.2 - "The Capital and Rental Valuation of Restaurants, Bars, Public Houses and Nightclubs in England and Wales" and at the time of compiling this chapter had been under review. The new version has now been published as a RICS Guidance Note (GN67/2010) providing more detailed guidance on licensed premises.

These documents encapsulate the profits method of valuation which works on the assumption that the landlord and the tenant will split the net profit before rent, with the landlord taking a percentage as rent and the tenant taking the balance as profit after rent. A 50/50 split is common, although valuers can debate a lower or higher percentage of the 'divisible balance' as rent depending on the supply and demand for a specific property and other factors such as the lease terms.

The RICS definition of rental value is market rent (MR). When adopting the profits method of valuation, this is based on fair maintainable turnover (FMT) of the reasonably efficient operator (REO).

The concept of FMT is that the landlord should not be punished for having an under-performing tenant and the tenant should not be penalised for performing above the level that might be expected for a REO.

The profits method of valuation requires a relatively high level of subjective judgment by the valuer, based on their knowledge and experience of the relevant sector of the market backed up by the interpretation of available comparable evidence.

Freehold sales with 'vacant possession' will often be as a fully equipped trading entity or 'unit'. They will include the goodwill that runs with the property but not the personal goodwill of the seller. The liabilities taken on will include compliance with relevant legislation and regulations which vary according to the property type. Transfer of Undertakings (Protection of Employment) (TUPE) Regulations will mean that the rights of existing staff have to be dealt with as part of a sale.

The amount of detail required to undertake a profits method valuation will in practice be driven by the proportionate use of time which is an issue even when acting as an expert witness. Initial assessments can be undertaken using a shorthand method, whereas a valuer making a case to an arbitrator or tribunal will normally adopt a more detailed approach. If the valuer strays beyond what a prospective tenant would do, in preparing a business plan, they may have gone too far.

It is possible to check the calculations with reference to elements found in each property type, such as rent per cinema seat, rent per restaurant cover, rent per barrel of beer sold but the valuer has to be able to analyse comparables consistently in what are generally imperfect sub-markets. That is a challenging task.

## FMT

A main driver of the method is the assessment of FMT. This is the annual turnover net of VAT that a REO is capable of achieving without over-trading or under-trading.

An REO is a competent operator acting in an efficient manner. The FMT of the REO will exclude the personal goodwill which is above market expectations and would be lost if that operator moved on.

In assessing the FMT, the target audience has to be considered. The current trading style might be wrong for the area and consequently the unit is under-performing. In assessing the rental value, the valuer is entitled to assume that the unit is trading to its FMT potential.

The exterior appearance of an outlet may be poor and could be affecting trade. Repainting, new signage and planting need not be expensive and can improve trade. Landscaping of outside areas (e.g. decking in trade gardens of pubs) is more expensive but can dramatically increase trade. The valuer may assume that work not requiring a landlord's consent as 'tenants improvements' has been undertaken and make an allowance for the capital expenditure.

The interior layout is something that often cannot be changed. There may, however, be non-structural alterations which can be undertaken to increase trade and if the REO would do it, then it should be assumed that the current tenant would do so. Leases will often say that it is the tenant's responsibility to comply with statutory obligations. Any structural alterations, however, are likely to be tenant's improvements to be disregarded unless a capital contribution was obtained from the landlord or a rent free period was granted in lieu of a tenant's capital contribution. Cosmetic improvements are easier to deal with and it is appropriate to consider if trade is being affected by the appearance of the interior.

Whilst consideration of these factors and experience will lead the valuer to a FMT figure, it is useful, as with all valuations, to consider comparable evidence. The turnover of comparables can be used to check the assessment of FMT. They can also be analysed to produce some other key indicators.

Once a figure for FMT has been adopted, the remainder of a shadow profit and loss sheet can be compiled which ends with a fair maintainable operating profit (FMOP). This is net profit before rent, otherwise known as the divisible balance.

## Gross profitability

Between establishing the FMT and arriving at a FMOP, calculations have to be made to calculate gross profitability from which are deducted operating and wage costs.

Gross profitability will depend on the products being offered and the wholesale supply line, which could be tied under the terms of a lease or franchise, and the sustainable retail prices of the products in that location.

## Costs

Once the total likely gross profit figure is calculated, likely running costs have to be deducted to arrive at a FMOP.

The costs are:

- Operational costs including: energy costs; rates; insurance; repairs; accountancy; sundries.
- Staff wage costs, including holiday pay and National Insurance.
- Tenant's capital including working capital, fixtures and fittings and stock. Here, the aggregate capital cost is calculated and an interest rate is then applied to produce an annualised deduction. This is based on a reasonable rate of return and/or the cost of money.

Once the total gross profitability from FMT and the likely costs have been calculated, the following calculation can be made: Gross profit from FMT less Costs = FMOP.

## FMOP

On a freehold property which is operator owned, the FMOP can be capitalised to arrive at a capital value.

On a leasehold property, the FMOP is available to be split between the property landlord as rent and the tenant as net profit after rent. Hence, the other name for FMOP is the 'divisible balance'.

From the available FMOP, or 'divisible balance', the landlord would like to maximise the rent and the tenant would like to maximise the 'net profit after rent'. If demand is strong for a unit there is an argument that the tenant's bid would be higher than 50%, and if the lease is onerous and/or on a weak site or where management is difficult, there is an argument that the tenant should have more than 50% to reflect the risk (i.e. the tenant's % rental bid is lower).

The percentage split of the divisible balance depends on the supply and demand at the time of the type of property in question and the risk/reward relationship to the figures.

Net profit after rent is often referred to as EBITDA, but EBITDA stands for earnings before interest taxation depreciation and amortisation. As stated above, an annualised amount is usually deducted for tenant's capital to arrive at the divisible balance.

## Commentary on use of actual accounts

The shadow profit and loss calculations will follow a similar format to the actual profit and loss sheet of the type of business in question.

There is a distinction, however, between the actual accounts of the business and the valuer's calculations, which are in theory what the REO will achieve at the date of valuation. Although the history of the site will be a strong factor indicating future trading potential, the prospective purchaser and therefore the valuer will look forward, not back.

As stated previously, the adoption of the trading potential of the REO is used so that the property landlord is not penalised for the under-trading of a poor tenant in the form of a low rent, and the tenant is not penalised for over-trading in the form of a higher rent.

The valuer, when undertaking a valuation of an owner operated entity, or when acting for a tenant, should have access to the actual accounts. An assessment as to whether the actual operator is trading above or below the level of a REO has to be made. The actual trading level is always useful in assessing FMT but only if the valuer stands back, looks at the relevant factors and considers whether, given all the circumstances, the tenant is over-trading or under-trading. The valuer can do this based on his or her experience and trading figures from other trading entities in the same market or sub-market. The level of competition and the prices that are sustainable to customers are important factors.

## Freehold valuation and sales

Once figures for FMT and FMOP are calculated, it is possible to calculate rental or freehold values.

For freeholds, where it is an investment valuation, an investment yield can be applied in the normal way to a rental stream, although estimated rental value can be calculated first to establish whether the property is under or over-rented.

There may be an overbid from a multiple operator who may have economies of scale and/or may wish to tie the site for a wholesale income stream in addition to rent. A higher or lower yield than the rent may be applied to the wholesale income to reflect the state of the market and the possible volatility of product sales.

Where the purchaser is an owner occupier, they may pay a capital figure based on a multiplication of net profit after mortgage repayment costs which might be more than net profit after rent.

Purchasers will need to fund their purchase and lending institutions will normally require the accounts of the current operator to compare with the purchaser's own forecasts. Where lending is driven by existing accounts, the amount a purchaser will be able to pay will be linked to performance and linked to the actual accounts which may not be in line with the valuer's opinion of FMT and FMOP.

## Licensed property

Licensed property is perhaps the largest sector where the profits test method of valuation is applied.

There is a wide variety of types of licensed property with different markets and sub-markets.

The subjective nature of the valuation methodology has brought it under scrutiny from time to time. Where possible, it may be avoided such as town and city bars let as a shell where there is sufficient evidence of similar space let on a floor area basis. Where this is not possible, no one has yet successfully promoted a better method of assessing properties which are sold or let based on their trading potential.

Other valuation approaches may be applicable to either complement or replace the FMT/profits method. In strong city/town centre drinking circuits, 'superpubs' can be analysed on the basis of capacity and average spend. Also units in urban locations with an open (A4/A3) user clause may be compared using an overall or zoning approach. The revision of the Use Classes Order in March 2005 split Use Class A3 (food and drink) into:

A3: Restaurants.
A4: Drinking establishments.
A5: Takeaway.

It is possible to move from being a drinking establishment under Use Class A4 to a Restaurant under Use Class A3 without planning permission. However, to move from A3 to A4 requires planning consent.

The Licensing Act 2003 put in place the ability to trade longer, subject to the consent of the local authority. This Act, which has been fully in force since 24 November 2005, replaced the justices on-licence with a requirement for a premises licence for the property. In addition to the premises licence, a person wishing to operate licensed premises needs to have a personal licence and there also needs to be a designated premises supervisor.

In July 2007, smoking in all public spaces in England was banned, which meant that only public houses with a suitable outside area could accommodate smokers.

Public disorder is a political issue, with excess alcohol consumption often cited as the reason for trouble in urban centres. Under the Licensing Act 2003, the police have the power to close down venues.

For publicans, the legislative burden is high, with the minimum wage, paid holiday entitlement, the Disability Discrimination Act and asbestos audits/management being just some of the legislative and regulatory controls introduced in the recent past. They can all impact upon the ability of a pub or bar to generate a profit and, hence, pay rent – as too will the increasing fuel and energy costs currently being experienced.

Before the credit crunch, multiple operators competed for group deals and this resulted in an overbid or lotting premium where competing purchasers could see economies of scale. It is important to clearly state whether a valuation is being undertaken on the basis of the individual establishments or as a group deal.

Generally, at the time of writing this chapter, the market has been depressed due to the recession and the market difficulties have resulted in a large number of corporate failures. The market for licensed premises may bottom out, although at the time of writing the threat of a double-dip recession, and increased beer prices due to a VAT increase and increasing wheat prices may prevent a recovery for some time. Existing operators and new entrants are beginning to take advantage of lower purchase prices and acquire additional premises.

The sale of leasehold businesses has been more difficult with the pre-credit crunch multipliers of EBITDA, or expressed as a percentage of FMT, not being achieved. The entry costs for a new 10, 15 or 20-year lease is prohibitive where capital is scarce and more short-term tenancies of up to three to five years are being granted.

Freehold prices have been substantially reduced and at the bottom of the market. Closed and/or boarded up premises have limited demand other than for alternative use. Where town centres have been over-supplied with contemporary bars, the cost of the original fit-out may often not be recovered.

## Pubs

Before the Government intervened in the brewery industry in 1989, most pubs and bars were owned directly by breweries, who ran them either as managed houses or under traditional tenancies.

This vertical (wholesale and retail) integration was seen as detrimental to competition by the Monopolies and Mergers Commission Report, 'The Supply of Beer', presented in March 1989. As a result of this report, the Secretary of State for Industry introduced two 'Beer Orders': The Supply of Beer (Loan Ties, Licensed Premises and Wholesale Prices) Order 1989 and The Supply of Beer (Tied Estate) Order 1989. Each brewery owning more than 2,000 pubs was ordered to release half the excess over 2,000 establishments from tie by sale or to let by 1 November 1992. Consequently, pub ownership and brewing was split and new specialist property companies emerged known as 'PubCos'.

These new property-owning companies offered 20 to 30-year leases. The leases were of interest to potential tenants due to the prospect of assigning these leases in the future and also the comfort of knowing that tenants of licensed premises had been given security of tenure under the Landlord and Tenant Act. For the majority, these new leases were 'tied', with the property companies nominating the brewery (which had previously run the outlet as a managed house) as the suppliers of the beer under the tie.

A 'tie' is an obligation to purchase supplies from a nominated supplier. That nominated supplier does not usually pass on all the wholesale discounts they enjoy to the retailer (in this case the publican). They might pass on some of the discounts, but often the retailer cannot buy beer for the same price as retailers who are free of tie or 'free houses'.

In the early 1990s, Inntrepreneur was one of the first of the new PubCos. It held tied leases with Courage as the nominated supplier. Other companies followed with different suppliers (e.g. Laurel with Whitbread/Punch with Bass). The level of discount offered on the supplied products was contentious, and the matter was referred to the European Court of Justice (ECJ) due to concerns about the anti-competitive nature of the leases.

'Negative clearance' under article 81 of the Treaty of Rome was granted for all but one of the main new PubCo leases due to the level of discounts and 'counterveiling benefits' (e.g. training and support). The one company that was not granted 'negative clearance' was Inntrepreneur. This led to a challenge in the UK Courts in the case of *Crehan v Inntrepreneur Pub Co (CPC)* (2003) High Court EWHC 1510 (Ch); (2004) EWCA Civ 637; 2006 UKHL 38, which was heard by the House of Lords in the summer of 2006, and resulted in a decision which essentially held that as the market was not foreclosed, the tenants, who claimed they were disadvantaged by high rents and low discounts, were not entitled to damages under European competition law.

In theory, rents of tied houses should be less than those of free houses to compensate tenants for this competitive disadvantage. There is some credence for the argument that tied leases are attractive because, in theory, the rents are lower, reflecting the lower gross margin that the tenant can achieve. A prospective tenant is unlikely to choose a tied lease in preference to a free of tie lease, except on the issue of rent and premium paid. New entrants to the business do sign up to tied leases due to their apparent affordability and the lack of available or afford-able free of tie leases.

The PubCos have been under much criticism of the tied lease models they operate and were scrutinised by a Trade and Industry Committee under the Labour Government, and recommendations were made including the abolition of machine ties.

At the current time, new Codes of Practice are being introduced which includes upward/downward rent reviews. The provision of Special Commercial or Financial Advantage (SCORFAs) by the Pub-Cos to tied tenants is an element in supporting tenants but these may not be contractual. In landlord and tenant matters there is normally a hypothetical landlord and a hypothetical tenant. If the SCORFAs would disappear if the actual landlord sold the freehold, it is ques-tionable whether they should not form part of a rental valuation.

A further complication is that some PubCos will not rentalise the machine income where the machines are tied. In the actual market, the hypothetical tenant would be expected to take machine income into account when making their rental bid so it will be interesting to see how this works through in rent reviews and lease renewals in practice.

## Valuation: public houses and bars

There has been wide publicity about the number of failures and clo-sures of public houses from 2008 to 2010, resulting from the impact of the smoking ban introduced in 2007, followed by the credit crunch and the recession at a time when publicans were competing with cheap supermarket alcohol. The British Beer & Pub Association publishes statistics, and it might be that the market is bottoming out. Where public houses now have less competition they should be more viable.

During the past 20 years, since the introduction of the PubCo lease model, trends for the licensed retail market have overall been negative. While local factors and variations should be con-sidered, the statistics from the British Beer & Pub Association Sta-tistical Handbook show that beer consumption is down (the rise of food, wine and premium packaged spirits has compensated for this) and there has been a sharp decline in the percentage of on-sales compared to off-sales from off-licences, supermarkets and personal imports. Wine sales have shown signs of increas-ing where food sales have increased and/or female custom has increased. The shift in balance between product lines has an effect on gross profit levels as the gross profit on wine is greater than that on beer.

Competition and the trading style of the outlet are key. There is only so much spend available in each neighbourhood for a particular trad-ing style. Two pubs in a small village may impact each other's trade unless they differentiate. A town centre circuit may be oversubscribed with vertical drinking halls, but an entrepreneur might still find a niche to meet an untapped demand (e.g. a comedy venue).

Where the lease is free of tie, the appropriate bar tariff can be considered and the wholesale cost deducted, to arrive at gross profitability from wet sales. If the lease is tied, however, the level of discounts offered by the PubCo must be considered. Low levels of wholesale discounts can reduce the gross profitability from circa 60 to 65% free of tie to 50% to 55% or below (depending on the level of discounts offered) on tied premises.

Food has become more important to sustain trade as traditional pub drinkers become scarcer or choose to drink cheaper supermarket alcohol at home (whilst possibly smoking at the same time). Bar snacks are easy to serve to attract drinkers.

To attract new customers the food offered may need to be improved but that may require more equipment and more staff. Food turnover may increase but the net profit overall may not. Also, once a kitchen is 'skilled up' and a chef is employed, the reputation of the establishment may be built upon an individual who may subsequently leave.

Machine income is taken as net, so profitability is 100% or the percentage split agreed with the landlord if there is a machine tie in place. As previously stated, a PubCo may agree to ignore net machine income in the calculation of rent.

For accommodation, the cost of providing bed and breakfast can be small once the rooms are established, so gross profitability can be circa 80% and above, depending on the deduction of the cost of providing breakfast.

Door entry is 100% gross profit, but door-staff have to be paid, so increasing wage costs.

Entertainment costs adopted should be those necessary to sustain the FMT, such as live and recorded music costs and subsidised or free food for pool and darts teams. Events, such as the World Cup, can give a business a boost but providing premier league football to customers regularly may not be cost effective due to the high cost of satellite TV subscriptions.

Wages vary greatly depending on the trading style and the extent of the food offered but are usually in the range of between 13% to 25% of FMT.

A contentious area is the inclusion of a manager's wage in the wage costs. If you assume that the tenant is the lessee/operator enjoying the financial benefit of living on the premises, the PubCos will usually argue that they should be working for the net profit after rent and not for a working wage. The hypothetical tenant might bid on that basis, but it does put the tenant in the position of working long hours for a limited return, which can work out at a very low hourly rate for management.

The position becomes even more difficult where there is a couple running the establishment. It can be somewhat inequitable that that net profit should be split between two people without at least one of them taking out a wage first. In the case of *Brooker* v

*Unique Pub Properties Ltd* (Chancery Division, 16 August 2001), Judge Weeks QC ruled that it was not unreasonable for licensees to earn £20,000 a year and that the rent should drop if it was impossible to achieve this. In *Brooker v Unique Properties Ltd* (2009) PLSCS 269, the judge dropped the tenant's bid to a lower percentage of the FMOP to give the tenants a fair return.

### Example 11.1

Set out below is an example of a profits method rental calculation of a free of tie public house in a good suburban location in the South East. It has an assessed FMT barrelage of 300 barrels and a healthy bar snack food trade. There is one Amusement With Prizes (AWP) machine, a quiz machine and a pool table. Entertainment is provided once a week. There are two bed and breakfast rooms that are full Monday to Friday at £30 per night. Note that in this and other examples VAT is deducted at 17.5% and VAT is due to rise.

| WET TRADE | | | | | | |
|---|---|---|---|---|---|---|
| Based upon 300 Barrels (NB 280 pints per barrel assumed sold to allow for wastage/ullage) | | | | | | |
| | Pints/litres | Average price excl VAT per unit | Turnover | Gross profit% | Gross profit | |
| Beer & cider | 84000 | £2.70 pint | £226,800 | 63% | £142,884 | |
| Packaged beer | 2000 | £2.55 bottle | £18,564 | 63% | £11,695 | |
| Wines | 1000 | £2.75 175ml | £15,703 | 70% | £10,992 | |
| Spirits | 300 | £1.85 25ml | £22,200 | 68% | £15,096 | |
| PPS | 700 | £2.55 bottle | £6,497 | 65% | £4,223 | |
| Minerals | 2000 | £1.10 various | £14,652 | 68% | £9,963 | |
| SUB TOTAL WET | | | £304,416 | 64% | £194,853 | |
| | Numbers per week | Average spend excg VAT | | | | |
| DRY TRADE | 100 | £4.25 | £22,100 | 50% | £11.050 | |
| NET MACHINE INCOME | 200 | £1.00 | £10,400 | 100% | £10,400 | |
| ACCOMMODATION | 10 | £25.53 | £13,276 | 80% | £10,620 | |
| SUB TOTAL INCOME | | | £350,192 | 64.80% | | £226,923 |
| LESS COSTS | | | | | | |
| Establishment | | | 15% | | | £52,529 |
| Wages | | | 15% | | | £52,529 |
| Bands/entertainment | | £100 | | | | £5,200 |
| | | | Inventory | £20,000 | | |
| | | | 2 Weeks stock | £4,741 | | |
| | | | Working capital | £10,000 | | |
| TENANT'S CAPITAL | | | | £34,741 | 10% | £3,474 |
| TOTAL COSTS | | | | | 32.5% | £113,732 |
| FMOP (DIVISIBLE BALANCE) | | | | | | £113,191 |
| TENANT'S BID | | | Say | 50% | | £56,595 |

Whilst this shorthand profits method spreadsheet can attract criticism for taking simple percentages for operational costs and wages, a total cost of circa 32 to 33% would not be unreasonable for an outlet of this type.

To check the wages one could assume 14 shifts per week x average 2 persons per shift (in addition to tenant) at £5.50 per hour (minimum wage £5.80 22+ £4.83 18–21). Thus: 14 × 2 × 6 × £5.50 × 54 weeks (54 not 52 to allow for paid holiday entitlement) × 2 persons = £49,896 to which National Insurance has to be added.

If there was a significant food offer then the salary of a part-time or full-time chef would have to be an addition. The provision of food may sustain the wet trade, but will not necessarily increase the FMOP and therefore the value of the property interest. The assessment of food potential can be one of the most contentious areas. Would the introduction of a skilled kitchen to replace microwaved frozen food increase turnover, and would the additional wage costs make it sustainable and worthwhile?

As can be seen from the example, the FMT is made up of the various income streams from the outlet. It is usual to start with draught beer, assessing the likely barrelage throughput. Stocktaking reports, delivery notes (if available) and cellar counts will guide the valuer on the split of the existing trade, although the FMT calculation does assume that the operator is keeping up with current trends.

Bottled beer is more popular now at urban venues than it was 20 years ago. Premium packaged products sell at a higher price per volume (although not necessarily at a higher gross profit). An assessment of bottled beer is required with likely throughput added to the draught beer.

Increased sales of wines and spirits in recent years have gone some way to balance the decline in on-sales of beer. These, generally, give a higher gross profit margin. This is a good example of where the shorthand and longhand profits method diverge, with the valuer able to either just add a percentage of non-beer wet sales or line-by-line assess litre throughputs of wines and spirits or even individual products.

Minerals are also more popular, with products such as J2O and mineral water as an alternative to coke or tap water with ice! Non-alcoholic drinks are obviously more significant at child-friendly venues and rural establishments with a trade garden.

## Comparables and market information

The hierarchy of market evidence starts with open market lettings. New rents, paid without a premium over and above inventory value and stock at valuation, provide arguably the best evidence. However, no two outlets are the same (in contrast with two floors of offices or two adjoining units on a business park), hence the adoption of the profits method of valuation.

One of the reasons that comparables and the key indicators that can be derived from them should be treated with caution is that leases differ (e.g. is the machine income kept 100% by the tenant or is it shared 50/50 with the landlord?). Assumptions in rent review clauses also differ. Questions should be asked such as: Is the unit assumed to be fitted out ready for trading or as a shell? Are there authorised tenant's improvements to be disregarded?

Whether the valuer is looking at new lettings, a lease renewal, rent review negotiated settlement or an arbitration award, the following may be used as a check:

<u>Rent as a % of FMT</u>. This might be circa 10 to 14% for tied leases and circa 12 to 18% for free of tie leases. One problem with this

analysis is that new lettings might show a higher percentage rent to turnover, but the tenant is often paying a higher rent, as he or she believes that the unit has been under-trading and they can increase turnover. The turnover in the tenant's business plan and not the actual turnover can be used to analyse the tenant's bid as a percentage of trade.

Rent per barrel. This might for example be circa £120 per barrel for a traditional outlet with no food sales or around £200 per barrel for a 'gastro-pub' with 50% food sales in a high tariff area.

## Valuation: late bars/clubs

Urban bars with late licenses and nightclubs are a different sub-market to public houses.

In planning terms, nightclubs used to fall within Use Class D2 (assembly and leisure) but are now specified as sui generis. What makes an A4 drinking establishment with extended hours under the Licensing Act 2003 into a sui generis nightclub under the revised Use Classes Order is not entirely clear.

Under the Licensing Act 2003, 24-hour opening is a possibility, although in reality most outlets applying for extended hours applied for one hour earlier (10 am) and one to four hours later (12 midnight to 3 am). The new Coalition Government, however, may review the previous government's policy of 24-hour drinking.

The conditions of the licence and fire safety issues will restrict capacity, and nightclubs have been losing some of their competitive advantage to pubs and bars that will have opened later.

Although the FMT may be more than a public house, the FMOP can be less due to the high cost of providing entertainment and security. The cost of licensed door staff is significant following the introduction of the Security Industry Authority (SIA) registration of trained doormen. Door entry fees may apply for music or comedy events or after 10 pm.

The product mix of a young person's venue which opens later than a traditional pub will be different. Non-beer sales will usually be higher with a higher proportion of sales being premium packaged spirits compared to real ale, for example. The gross profitability on wet sales may be higher, although this is countered by price wars where competition is high, and wet sales can be significantly affected by customers drinking alcohol before arrival "pre-loading" and then drinking water which has to be provided in club venues.

### Example 11.2

This shows an increase of just above 50% in FMT to reflect the extra capacity and hours (no accommodation) but as can be seen, the extra cost of entertainment and security, results in a rent only circa 15% higher than in Example 1.

| WET TRADE Based Upon 400 Barrels | | | | | | |
|---|---|---|---|---|---|---|
| | Pints/ litres | Average price exc VAT per unit | Turnover | Gross profit% | Gross profit | |
| Beer and cider | 112000 | £2.70 pint | £302,400 | 63% | £190,512 | |
| Packaged beer | 4000 | £2.55 bottle | £37,128 | 63% | £23,391 | |
| Wines | 3000 | £2.75 175ml | £47,108 | 70% | £32,975 | |
| Spirits | 600 | £1.85 25ml | £44,400 | 68% | £30,192 | |
| PPS | 5000 | £2.55 bottle | £46,410 | 65% | £30,167 | |
| Minerals | 4000 | £1.10 various | £29,304 | 68% | £19,927 | |
| SUB TOTAL WET | | | £506,750 | 64.56% | £327,164 | |
| | Numbers per week | Average spend exc VAT | | | | |
| DRY TRADE | 100 | £4.25 | £22,100 | 50% | £11,050 | |
| NET MACHINE INCOME | 200 | £1.00 | £10,400 | 100% | £10,400 | |
| ACCOMMODATION | 0 | £0.00 | £0 | 0% | £0 | |
| SUB TOTAL INCOME | | | £539,250 | 64.65% | | £348,614 |
| LESS COSTS | | | | | | |
| ESTABLISHMENT | | | 15% | | | £80,887 |
| WSEGA | | | 15% | | | £80,887 |
| ENTERTAINMENT | | | £500 | | | £26,000 |
| SECURITY | 24 | £20 | £480 | | | £24,960 |
| | | | INVENTORY | £30,000 | | |
| | | | 2 WEEKS STOCK | £7,332 | | |
| | | | WORKING CAPITAL | £10,000 | | |
| TENANT'S CAPITAL | | | | £47,332 | 10% | £4,733 |
| TOTAL COSTS | | | | | 40.3% | £217,467 |
| FMOP (DIVISIBLE BALANCE) | | | | | | £131,147 |
| TENANT'S BID | | | Say | 50% | = | £65,573 |

The risk/reward ratio on a young person's venue might result in a higher or lower % rental bid. If this was a bar or club in an urban centre it would be usual to be applying rents per square metre (foot) as there would probably be sufficient comparable evidence to do so (with appropriate weighting of space not on the ground floor or ancillary).

# Restaurants

Restaurants, by contrast, are typically compared using either an overall (per square metre/foot) or covers basis. In neighbourhood parades, it is common to still find valuers applying the retail tone In Terms of Zone A (ITZA) to a restaurant and adding a percentage for an A3 use. When A3 use was first introduced, valuers got drawn into the habit of adding 10% for A3 use over A1 use values in a parade and this is something still seen today. This may ignore a higher level of rental demand for restaurant units in the wider local area.

There is also a planning distinction now for takeaway use, which has its own use class; A5. In some areas it is possible to find that A5 rents are 30%, or more, higher than retail rents.

Valuers should, if possible, try to find comparable evidence of similar restaurants in an area wider than the subject parade and an open market sale or letting will be at the top of the hierarchy of evidence.

Where a valuer is not able to rely on an overall rent (per square metre/foot), rent per cover or zoning methodology from comparable evidence, they should resort to a calculation of rental value from estimated profit and turnover. This mirrors how an operator, in theory and often in practice, will assess a rental bid in the marketplace.

The level of competition and the potential turnover and profit of similar restaurants in similar locations is relevant, although a multiple operator will see the presence of other restaurants as positive in a lot of cases. The branding of restaurants makes the disregard of goodwill generated by the brand a live issue, but only where another brand in the market could not trade at the same level in the same location.

Gross turnover net of VAT can be calculated using the following formula:

Covers × sittings × average net spend excluding VAT × days.

Gross profit from this trading potential depends on the style of food and can vary widely. Unlike public houses, there are no tied businesses other than franchises where a brand is operated locally by the individual owner.

Operating costs and wage costs will be deducted, as per the public house example, to arrive at FMOP. An annualised amount for tenant's capital should also be deducted. Careful attention must be given to the treatment of fit-out costs. A lot of restaurants are let on a shell basis.

There are higher staff costs in restaurants where the kitchen is skilled and table service is required.

As with wet led establishments, there is a level of subjectivity in this valuation approach which can be tested by analysing comparable evidence. As with all valuations, if there is no reliable comparable evidence, the level of experience the valuer has will carry weight.

---

**Example 11.3**

Where a profits method approach is adopted, the following is an abbreviated example of a 40 cover restaurant in a large affluent village with no competition:

| | | |
|---|---|---|
| Monday (exc Bk hols) – Thurs | 20 covers × 1 sitting × £10 × 200 days | £40,000 |
| Fridays (exc Good Friday) | 30 covers × 2 sittings × £15 × 51 days | £45,900 |
| Weekends and Bank hols | 40 covers × 2.5 sittings × £20 × 114 days | £228,000 |
| Total | | £313,900 |
| Gross profit say 65% | | × .65 |
| | | £204,035 |
| Operating costs say 16% | = £50,224 | |
| Wage costs at 25% | = £78,475 | |
| Tenant's capital £100,000 at 10% | = £10,000 | |
| Total costs | | £138,699 |
| FMOP (divisible balance) | | £65,336 |

At a 50% tenant's bid, the rental value would be say £32,500 per annum which is marginally above 10% of FMT.

This example shows £100,000 tenant's capital, but rents as a percentage of FMT will often be lower in a restaurant than for a public house due to higher wage costs based on food preparation and table service.

# Hotels

The hotel market is a volatile one, significantly affected by outside events. In the short term this might be a health scare or volcanic ash. In the medium term, the economy, foreign currency trends and holiday habits are some of the factors.

As with other classes of leisure property there are sub-markets within the hotel sector. The number of branded budget hotels in towns and cities and alongside main roads has increased, whilst the viability of traditional and older two or three star hotels has been affected. The market has become more polarised with the creation of more contemporary boutique hotels at its upper end.

In the provinces, many pubs have developed letting rooms. Golf, resort and spa facilities have been added to hotels to extend their appeal. There are luxury and country house hotels along with boutique hotels (often bijou town houses) at the top end, and at the bottom there are budget and lodge hotels, which compete with independently owned small hotels and bed and breakfast establishments. Airport and seaside hotels have their own patterns of trade, the former with high occupancy and the latter being seasonal. Aparthotels (extended stay) has been an emerging market.

The types of tenure are as follows:

- Freehold vacant possession, i.e. owned and operated by the same person.
- Owned but operated by a tenant.
- Owned but operated under management.
- Owned but operated under franchise.

The RICS Valuation Paper No 6: 'The Capital and Rental Valuation of Hotels in the UK' promotes the 'income capitalisation method' for the majority of mid range hotels which are to be valued with vacant possession. This method is used in conjunction with the discounted cash flow (DCF) method for more upmarket hotels in the four and five star categories.

Branded and/or budget hotels are often run on a franchise basis or subject to a management agreement. These arrangements need to be fully understood by the valuer in a hotel sub-market.

The valuer has to be aware of the 'lifestyle' purchaser at the lower and/or more remote end of the market, who may see the hotel as a 'trophy'.

Valuations and market transactions are, however, on the whole, based upon net cash flow. The valuer's role is to assess the trading potential, work out the profit it can generate and apply a multiple to that income with any adjustments for necessary or desired capital expenditure.

RevPAR (Revenue per Available Room) is a key indicator of performance which is calculated by multiplying occupancy x ADR (average daily rate or average room rate). Occupancy is the percentage of available room-nights occupied. The yield per room/yield per bed can be calculated and compared with other hotels.

The potential income streams, with different gross profitability levels, are split. These are as follows:

- Accommodation
- Food
- Drink
- Sundries
- Any on-site leisure spend (e.g. spa treatments).

A figure is arrived at which, when valuing hotels, is termed income before fixed charges (IBFC).

Operating costs and wages (also tenant's capital if appropriate) are deducted as per other classes of property with operating costs tending to be higher and wages close to restaurant percentages.

Depending on whether a rental value or capital value is being derived, a split of the remaining balance or a multiple of net operating profit can be applied.

A DCF technique can be criticised as being a technique for assessing worth to a specific individual rather than assessing market value. By adopting FMT and not actual turnover, the valuer can avoid the accusation that he or she is valuing worth to an individual operator.

With a DCF exercise, the normal time-frame to adopt is five or 10 years, but that requires an assumption to be made about the reversion at the end of year five or 10.

The analysis of comparables, ideally recent market transactions, is important. A capital sum or rent per room may be used as a check with reference to comparables but used with caution unless there are a number of similar hotels in the same area (e.g. a seaside town).

## Other types of property

The other types of property where the profits test method of valuation can be applied, are quite diverse.

Urban Leisure is a phrase used to cover such venues as cinemas, theatres, bingo halls and ten pin bowling.

It may also cover nightclubs, which have already been dealt with in this chapter.

Below is a brief resume of a few other types of property where the profits method is applicable, although this is not exhaustive. Each could probably warrant a chapter in their own right. There is not room in this publication to provide working examples as to how these properties are dealt with, but the author hopes that the following will give the reader a flavour of what is involved, and inspire further investigation. There is not the space in this publication to even introduce some sub-markets such as gymnasia and sports facilities, caravan parks, theme parks, visitor attractions and casinos, all of which are highly specialised.

## Cinemas and theatres

Cinemas can follow a similar format to the licensed leisure sector in terms of working out FMT and then deducting costs, which includes film hire, to arrive at a FMOP. The income streams are divided between admissions and confectionary sales, with additions for screen advertising and car parking if appropriate.

Where there is sufficient comparable evidence (e.g. in the mature multiplex market), calculations can be based on rent per square metre (foot). Check calculations based on rent per annum per seat can be of assistance, but there has to be awareness that the different style of auditoria can result in widely differing operating costs. For example, an 'art' cinema may fall into a different sub-market.

The multiplex market has matured and older style cinemas are becoming rarer with many being converted to bingo halls and 'superpubs'.

Other than in London and some other major cities, theatres may not be profitable. Smaller provincial theatres are often owned or run by trusts. The appropriate valuation approach may be Depreciated Replacement Cost (DRC) if there is no commercial market for the building.

## Care homes

Care homes are also valued with reference to trading potential. The valuer has to be aware of the different regulations covering different care situations. Like hotels, calculations based upon rent per room and DCF calculations can be used.

Capacity, occupancy rate and average fee rates have been used to calculate FMT.

Specialist care homes with the need for more acute care can result in higher revenues but the costs of providing care will inevitably be higher.

## Petrol filling stations

Petrol filling stations are assessed based on throughput per litre/gallon. The key is establishing 'genuine' sustainable trade.

Ties to petrol companies and franchise agreements are common, but value will be driven by whether or not which operators will be interested in the site with major oil companies having a minimum throughput threshold. Petrol forecourts at superstores have affected the viability of older roadside stations.

Shop sales have become very significant over the past 10 to 15 years, with partnerships developed between petrol retailers and supermarket operators. Care needs to be taken to identify income streams and the rights and liabilities that run with the property interest being valued.

## Golf courses

These can be assessed based on the grade of course, membership numbers, the golf and ancillary income.

Since the golf development boom of the late 1980s and early 1990s, the bottom end of the market has become oversupplied in a lot of areas, but top end courses can still command good demand from investors and operators.

Established member-owned clubs built many decades ago are rarely sold, and can command premium values. These are run for the benefit of members with the target being a surplus each year to cover major expenditure in the future. They compete with commercially run courses although they are run on a different basis.

## Racecourses, racetracks and stadia

Other outdoor venues can be evaluated by considering the status of the venue within the sector, the sustainability of events and income streams.

A particular venue may have a unique value which is difficult to quantify, but the valuer needs to comment that a premium value would be likely should the property ever be sold.

Stadia are generally attached to a team. Whilst the ownership can often be separated from the ownership of the team whose home venue it is, the base value is driven by sustainable sporting revenue (which can be volatile depending on team performance over time). The stadium, can, however, be seen as a trophy in its own right and ancillary accommodation such as conferencing facilities can add value over and above a capital amount for the pitch, seating and concourses.

## Tenant's improvements under the profits method

Tenant's improvements in mainstream commercial property, such as offices, are generally more straightforward to deal with than in the property sectors where the profits method is applied. The general rule under the Landlord and Tenant Act 1954 and the Law of Property Act 1969 is that authorised tenant's improvements are not rentalised for 21 years (in effect the next rent review or lease renewal after 21 years).

There are two schools of thought on the disregarding of tenant's improvements when adopting the profits method. The first is, in line with mainstream commercial property, that you value what was there prior to the tenant's improvements being undertaken.

The second approach is to the value on the basis of the trading potential with the improvements but deduct an annualised amount to reflect the capital costs of the improvements at the valuation date.

Dual rate can be used to annualise the capital cost (at the valuation date, not historic cost) of the tenant's improvements, adopting the hypothetical term in the lease (bearing in mind landlord and tenant rights to renew) or a shorter period (e.g., 10 years) reflecting the convention of the sub-market. In the example below a single rate is used in perpetuity for illustrative purposes only.

With trading related property it is important to remember that an individual's additions or alterations to a property may not increase the FMT and/or the FMOP. There are good improvements and bad improvements. The REO should undertake improvements which will provide a return on the investment.

Where comparables are available it is possible to assess the property prior to the tenant's improvements.

Where comparables are not available, however, the 'top down' approach may usually be adopted. The 'top down' approach is to assess the trading potential with the improvements and then deduct an annualised amount for the capital cost of building those improvements at the valuation date.

The author of this chapter is not promoting the 'top down' approach as the primary method. There may be case law in the future which gives valuers more guidance. The law on improvements, however, is complex and not always conclusive with respect to an individual case the valuer is dealing with.

It is important to remember that there should be only one rental value for the property, so by using both approaches and cross referencing the valuer should, in theory, arrive at the same figure from both approaches. This might be achieved by adjusting the tenant's percentage rental bid of the divisible balance to reflect the risks/rewards of carrying out or not carrying out the tenant's improvements.

An example of the 'top down' approach would be as follows:

### Example 11. 4

| FMT | £350,000 | | | | |
|---|---|---|---|---|---|
| GROSS PROFIT | | | | | £225,000 |
| LESS COSTS | | | | | |
| ESTABLISHMENT | | 15% | | | £52,500 |
| WAGES | | 15% | | | £52,500 |
| ENTERTAINMENT | £100 | | | | £5,200 |
| | | INVENTORY | £20,000 | | |
| | | 2 WEEKS STOCK | £5,000 | | |
| | | WORKING CAPITAL | £10,000 | | |
| TENANT'S CAPITAL | | | £35,000 | 10% | £3,500 |
| TENANT'S IMPROVEMENTS | | | £100,000 | 10% | £10,000 |
| TOTAL COSTS | | | | | £123,700 |
| FMOP (DIVISIBLE BALANCE) | | | | | £101,300 |
| TENANT'S BID | | Say | 45% | = | £45,585 |

The example shows the custom of making an annualised deduction for the capital cost of the improvements as an addition to the overall costs to be deducted from gross profit to arrive at a divisible balance.

Placing the deduction in at this point (rather than simply as an end allowance) means that, without other items being adjusted, any increase in gross profit as a result of the tenant's improvements, which is higher or lower than the annualised amount for tenant's improvements, will have an impact on the figure on the divisible balance line and therefore the rent, unless the percentage tenant's bid is adjusted. (The author acknowledges that this could also be done by adjusting the yield applied to the annualisation of the tenant's improvements but generally valuers do not appear to do that.)

The rent might, therefore, be higher with the tenant's improvements, which should have been disregarded. The tenant might be compensated for this by receiving additional profit after rent left from the divisible balance after the rent has been adjusted. Where the landlord has contributed to the costs of the scheme which has resulted in a higher gross profit, it may be fair that a higher rent results.

Where the tenant's improvements to the building have actually had a negative impact on gross profit, this model should show that the rent is not increased. You could run the figures, taking £200,000 as gross profit and the divisible balance would drop to £76,300 and at a tenant's bid of 45% that is a rent of £34,335. If this gross profit would have been achieved without the tenant's improvements being undertaken, removing the tenant's improvements line would increase the divisible balance to £86,300 which, at a 50% tenant's bid on an unimproved bottom up approach, would produce a rent of £43,150.

The reader may like to try other permutations with different gross profit figures and different annualised amounts for tenant's improvements to see how sensitive the model is, and why careful consideration of the percentage tenant's bid and a stand back approach at the end is advisable.

Comparables, if available, will of course assist. In Example 11.4, if there are comparables on a 'bottom up' approach, which suggest that the unimproved property should have a rent of around £45,500, the valuer can stand back and be satisfied that the 'top down' approach has produced a figure at the right level.

## Summary

The valuer must have knowledge and experience of the market or sub-market in which the property rests in order to be able to apply knowledge and experience to the subjective judgments that have to be taken.

Given the possible combinations of trading style, lease and use, the profits method provides a way of assessing the rental value of each outlet on its potential merits and has much to offer – but relies heavily on the expertise and skill of the valuer.

Comparables, as with all property sectors, are important. The valuer needs to know when and how to adopt the figures based on data from comparables. The valuer can adopt key market ratios with assistance from benchmarking sources.

To conclude, it is worth noting that the nature of the relationship between the landlord and the tenant is more complicated where the profits method is applied than in the mainstream commercial sector (e.g. shops, offices) as the existence or absence of supply ties complicates the market.

It is more difficult to apply objective modelling to valuations, particularly as the extent of ties differs; furthermore every establishment is individual with different pricing and business potential.

Although the author has limited references in the text, the bibliography contains publications that are recommended reading for those who wish to obtain further knowledge.

The profits method is a challenge but an enjoyable one for those readers who wish to develop their valuation skills beyond the overall (per square metre/foot) method.

# Chapter 12

## Investment Analysis

## Introduction and re-cap

The growth in institutional investment in land and buildings in the 1970s provided an impetus for valuers to think about the way in which their valuation approaches related to more widely-used methods of investment analysis, especially those used for appraising the stocks and bonds held by those investors. Much was written about 'rational' valuation methods, especially those based on discounted cash flow. Nevertheless, the violent boom and slump of property prices in the early 1970s, the late 1980s and 2007–09 suggests that the market still had much to learn about the relationship between economic activity and the property market.

When property is purchased as an investment, it is purchased for its present and future income and for capital growth. Occasionally, other factors (such as prestige) enter into the decision, but for the rational investor such factors should not override decisions based on sound appraisal. This is by no means uncontroversial, but most market participants would expect that the valuer should take part in this process and not merely 'keep the score' by observing market behaviour. This can be argued if only to make the point that it is likely to be useful for the valuer to see an investment from an investor's perspective in order to inform the valuation. This requires careful consideration of expected returns, which can only be estimated by a consideration of the future, which is unknown. How can we start to look into the mind of an investor, and how might the investor think about the expected return?

The following is an example press comment from the *Financial Times* in early 2010:

> London's commercial real estate market is thriving again, and the Square Mile's tallest worked-in skyscraper, Tower 42, is to be put up for sale, The Financial Times reports. About £300 million ($461 million) is expected when Hermes Real Estate and BlackRock UK's real estate fund, the owners of the 600-foot tower, open it up to bidders this week, the paper says.
>
> The sale of the tower will be the largest single real estate deal in the City this year, and indicates the recent growth of interest among buyers for prize assets.

The current weakness of the pound, coupled with a predicted scarcity of prime office space, has led to a surprise boom in the London market, with the prices of certain buildings now close to what they were at its peak three years ago, The FT says. Tower 42 was considered Britain's first true skyscraper when it was built for NatWest 20 years ago. It generates a rental income of around £20 million a year.

In this comment, we can quickly see the simple relationship between rental value and capital value. A price of £300 m is expected, and the property generates a rental income of around £20 million a year. This will provide an initial yield of 6.67%, and (looking at this the other way round) the valuation has been arrived at using a capitalisation rate of 6.67%, so that £20 m/0.0667 = £300 m. However, this tells us nothing about the expected returns, which can only be estimated by a consideration or projection of the future.

The following is another example taken from a website:

Light Industrial, Office - for sale, to let
Unit 21, Kings Meadow, Ferry Hinksey Road (off Botley Road)
Oxford
A modern two storey business unit, gas heating, flexible ground floor use with loading door, parking. Please note that this property is also available freehold.
2,504 sq ft
Rent: £17,500 pa (£7 psf)
Price: £220,000 (£88 psf)
Tenure: Freehold, Leasehold

In this case, a price of £220,000 is expected, and the advertisement suggests that the property (albeit vacant) could generate a rental income of around £17,500 a year. This will provide an initial yield of around 8%. Looking at this the other way round again, the valuation has been arrived at using a capitalisation rate of 8%, so that £17,500/0.08 = c. £220,000. This simple relationship is expressed as shown in Chapter 5 and as follows:

rent
*multiplied by*
cap factor (present value (PV) of £1 pa in perpetuity)
*equals*
capital value.

In this case the rent is found from the size of the lettable area and comparable evidence of rental values of similar property per unit of space, making adjustments for perceived differences in quality. The calculation is as follows:

$$2,500 \text{ sq ft} \times £7 \text{ psf} = £17,500$$

The capitalisation factor is also found from comparable evidence. Someone (a valuer or the broker) has estimated that other similar properties have sold for around 12.5 times rent, perhaps making adjustments again for perceived differences in quality. The valuation would look like this:

| | |
|---|---|
| Market rent | £ 17,500 |
| PV £1 pa in perpetuity at 8.00 % | × 12.50 |
| Capital value : | £ 218,750 |
| Say : | £ 220,000 |

This can also be arrived at quickly by £17,500/.08 = (say) £220,000. The initial yield (income return): £17,500/£220,000 = 8%, ignoring purchaser's costs for simplicity.

However, as we discussed in Chapters 5 and 6, it is not common practice to ignore the purchaser's costs in such a case. These need to be deducted from the total cost to leave the amount available for the property acquisition. Assuming that the purchaser's costs are 5.75%, the maximum price will then be £218,750/1.0575 = £206,856.

What does this tell us about the expected return? We are provided with very little information – simply that the purchaser will receive an initial yield of 8% on his or her total outlay. Whether that income yield will rise or fall, or whether the property will grow in capital value or not, and what the expected return or internal rate of return (IRR) will be over a given holding period, are all unknown – or, more likely, hidden assumptions.

## Expected returns – the cash flow

Expected returns can only be estimated by a consideration of the future, which is unknown. So in order to speculate about expected returns we will have to make some reasonable assumptions, primarily the expected cash flow. As we saw in Chapter 3, the process of discounted cash flow underpins any analysis of value. The appraisal of all investments is predicated on the assumption that the current value is equal to the net present value (PV) of the future benefits (see, for example, Damodaran (2001) and, in a property context, Baum and Crosby (2008)). This requires us to determine the most likely cash flow that the investment will produce, and the discount rate which we can then use to find the net present value of that cash flow.

Many issues arise when constructing a property cash flow. The generation of some of these inputs can be challenging and complex, and there is a full menu of issues to consider would be very long and very detailed, as set out in Chapter 3 of Baum and Crosby (2008).

For simplicity, let us assume that the property will be let quickly at an initial rent of £17,500 net (on full repairing and insuring (FRI) terms) so that this represents what is known as the net operating income (NOI) on a new five-year lease at a fixed rent. Assume that the lease is renewed in years five and 10, and that we sell the property in year 10, having agreed a new rent, at the same capitalisation rate. Assume rent is received annually in arrears and, an important assumption, that rental values will rise at 1.75% pa. This is the expected inflation rate at the time, and there is strong evidence that rental values follow inflation in the long run (see, for example, Baum (2009)).

If this rental increase is achieved at review, the rent agreed in year 6 will be:

$$(£\,17,500 \times (1.0175)^5) = £\,19,097$$

and the rent agreed in year 11 will be:

$$(£\,17,500 \times (1.0175)^{10}) = £\,20,817$$

This produces the following cash flow for years 1 to 15:

| Year | Rent |
| --- | --- |
| 1–5 | £17,500 |
| 6–10 | £19,087 |
| 11–15 | £20,817 |

If we sell the building at the end of year 10 at the same cap rate, we will receive £20,817/.08 = £260,216. Our cash flow then looks like this:

| Year | Rent |
| --- | --- |
| 1–5 | £17,500 |
| 6–9 | £19,087 |
| 10 | £19,087 + £260,216 = £279,303 |

## The discount rate

Next, we have to determine a discount rate. There are two very different ways to approach this. First, we could ask: what target or hurdle rate would the typical buyer of this property use? Alternatively, we could estimate what hurdle rate we believe to be appropriate (see later).

Fortunately, by making one key assumption, we have all the information we need to be able to precisely calculate the target or hurdle rate we can assume that the typical buyer of this property would use. This derives from the following simple equation:

$$K = R - G \qquad (1)$$

where:

- K is a capitalisation rate.
- R is the required return or discount rate.
- G is growth in rents.

Further:

$$R = RFR + RP \qquad (2)$$

where:

- RFR is a risk free rate, such as a bond yield (see Chapter 6).
- RP is a risk premium, an extra return to compensate for risk.

Combining equations (1) and (2), we can derive equation (3):

$$K = RFR + RP - G \qquad (3)$$

Taking equation (1):

$$K = R - G$$

In this case:

- $8\% = R\% - 1.75\%$.
- $R = 9.75\%$.

(Note that the key assumption we have made is that the typical buyer of this property would assume that rental values will rise in line with our own assumption, 1.75% or the rate of inflation.)

This estimate of a required return or discount rate of 9.75% is an approximation. Equation (1) assumes annual income increases. However, this property is subject to five-yearly reviews. This means that we need more annual rental growth, or a lower discount rate, to compensate for this.

It can be shown that a capitalisation rate of 8% gives exactly the same result as if we assume growth of 1.75% and a discount rate of 9.5%. From Baum and Crosby (2008), equation (1) should be modified to equation (4), shown below.

$$K = R - (SF \times P) \qquad (4)$$

where:

SF = the annual sinking fund for N years at R for the review period.
P = % growth in rents expected over the market review period.
N = the rent review period (five years in this case).
$SF = R/[(1+R)^{\wedge}N - 1]$.

where

$$G = 1.75 \div, P = (1+0.0175)^5 - 1 = 0.090617$$

By iteration, where $K = 8\%$, $R = 9.5\%$. Hence, the projected ERV (estimated rental value) in year 6 is $= £17,500 \times (1+.0175)^5 = £19,087$. The full annual cash flow, discounted at 9.5%, is as shown below.

| Year | Rent | Capital | PV factor at 9.5% | PV |
|------|------|---------|-------------------|-----|
| 2010 | £17,500 | | 0.9132 | £15,982 |
| 2011 | £17,500 | | 0.8340 | £14,595 |
| 2012 | £17,500 | | 0.7617 | £13,329 |
| 2013 | £17,500 | | 0.6956 | £12,173 |
| 2014 | £17,500 | | 0.6352 | £11,116 |

*(Continued)*

| Year | Rent | Capital | PV factor at 9.5% | PV |
|---|---|---|---|---|
| 2015 | £19,087 | | 0.5801 | £11,073 |
| 2016 | £19,087 | | 0.5298 | £10,112 |
| 2017 | £19,087 | | 0.4838 | £9,235 |
| 2018 | £19,087 | | 0.4418 | £8,433 |
| 2019 | £19,087 | £260,216* | 0.4035 | £112,703 |
| * £20,817/.08 | | | | £218,750 |

Looking at the cash flows lease by lease and capitalising the final cash flow in perpetuity at the capitalisation rate, the same result is obtained.

| Year | Rent | PV £1 pa | PV | CV |
|---|---|---|---|---|
| 1–5 | £17,500 | 3.8397 | 1.0000 | £67,195 |
| 6–10 | £19,087 | 3.8397 | 0.6352 | £46,554 |
| 11– | £20,817 | 12.5000 | 0.4035 | £105,001 |
| | | | | £218,750 |

Note that the PV of the cash flow is £218,750, exactly the same result given by £17,500/.08. This is simply a more informative analysis of market values, working from the same assumptions, but breaking the capitalisation rate (K) of 8% into its two components of required return (R) and growth (G).

Note also that, given a market observation of K, the calculation of one variable R or G will depend on the assumed value of the other. This assumed value will be immaterial to the result, because a high value for G will be compensated by a high value for R, and vice versa, and the value will not change.

## Some simple analytical measures

Some reference has already been made to problems of terminology relating to yields (see Chapter 4). As already suggested, there are many confusions concerning return measurement in property. This is largely due to the unique terminology which has grown in the property world; it is also due to the unique nature of property, especially its rent review pattern and the resulting reversionary or over-rented nature of interests.

There is also some misunderstanding of the difference between return measures which are used to cover different points or periods in time. Return measures may describe the future; they may describe the present; or they may describe the past.

Measures describing the future are always expectations. They will cover periods of time and may, if that period begins immediately, be called *ex ante* measures. An example is the expected IRR from a property development project beginning shortly; another example is the required return on that project.

Measures describing the present do not cover a period, but describe relationships existing at a single point in time (now). An example is the initial yield on a property investment; this is simply the current relationship between the rental income and the capital value or price.

Finally, measures of return describing the past, or *ex post* measures, are measures of historic performance. An example is the delivered return on a project.

The following definitions describe the present.

## Initial yield

This is defined as the net rental income divided by the current value or purchase price. Similar concepts are used in other investment markets; these include interest yield, running yield, income yield, flat yield, and dividend yield.

The initial yield may mean the relationship between the before tax net rents receivable during the first 12 months of ownership and total acquisition cost; the return expected during the investor's current financial year; or the relationship between current contracted rents and total acquisition cost. In the absence of specific direction from the client it would be normal to assess the relationship between current contracted rent and total acquisition costs.

In Chapter 3 the valuation of a property let at £10,000 per annum for two years, with a current market rental value of £20,000 per annum, was considered in some depth. The relationship between the all risks yield (or market capitalisation rate) of 5% and the required return of 10% was examined, and on a five year rent review pattern an implied annual rental growth of 7.6355% was calculated. This produced an implied rent in two years of £23,170, and a capital value on a modified DCF basis of £386,323.

For example, let us assume the property is acquired for a price just below valuation of £378,100. Adding legal fees (1%), surveyor's fees (0.75%) and stamp duty (4%) at a total of 5.75% of the purchase price, the total outlay is £378,100 × (1.0575) = £399,844.

| | |
|---|---|
| Current gross rent | £10,000 |
| Less landlord's expected non-recoverable expenses | £500 |
| Net rent | £9,500 |

Initial yield: £9,500/£399,844 = 2.376%

This yield can be crucial to some investors. Equity investors – those using their own capital to purchase, such as insurance funds and pension funds – may need a minimum income yield to make a property asset attractive relative to cash or bonds. Debt investors (those borrowing a large proportion of the acquisition costs) will need a minimum income return to pay interest charges.

Other funds, while having regard to the initial yield and its implication for their whole portfolio, will be aware that a low initial property yield can be balanced by income growth to achieve a total return for the fund above the target rate. For these investors, initial yield is less vital.

## Yield on reversion

This is defined as the current net rental value divided by the current value or purchase price. There is no equivalent in other markets.

Investors using debt may be able to finance a shortfall in rent for a period of time, but when holding reversionary property will be interested to know whether, on the basis of reasonable or conservative assumptions, they will be able to cover interest charges when the reversion is due and a full market rent is paid. It is therefore normal in an investment report to include an assessment of the yield at reversion calculated on the basis of current estimates of open market rentals.

| | |
|---|---|
| Estimated rental value | £20,000 |
| Less non-recoverable management fees say | £750 |
| Net rent on reversion | £19,250 |

Yield on reversion: £19,250/£399,844 = 4.814%.

## Equivalent yield

This is the average of the initial yield and the yield on reversion. It can be defined as the IRR that would be delivered assuming no change in rental value. As in the case of all IRRs, the solution is found by trial and error. There is no equivalent in other markets. Many investors find comfort in this conservative or risk-averse way of looking at the investment, which can also be thought of as an average of initial and reversionary yields.

What is the expected cash flow assuming that rental values and capitalisation rates do not change? Let us assume that we sell the building at the reversion. Then we need to estimate the resale value.

| | |
|---|---|
| Rental value | £20,000 |
| PV £1 pa in perp at 5% | 20 |
| Capital value | £400,000 |

Then the IRR on an annual in arrears assumption is found from the following cash flow.

| Year | Cash flow |
|---|---|
| 0 | − £399,884 |
| 1 | + £9,500 |
| 2 | + £409,500 |

The IRR of this cash flow, found by trial and error, is 2.395%. This is the equivalent yield, lying somewhere between the initial yield of 2.376% and the yield on reversion of 4.814%. (If acquisition costs are excluded from this analysis, the equivalent yield will be closer to the yield on reversion.)

An alternative approach to estimation of the equivalent yield is to use trial and error applied to a term and reversion valuation with a single, unknown yield throughout. The yield or capitalisation rate which produces the capital value of the investment given the current rent, the current rental value and the term to reversion is the equivalent yield.

## Reversionary potential

This is the net rental income divided by the current net rental value or vice versa. There is no equivalent in other markets.

The following definitions are performance measures. They describe the past.

## Income return

This is the net rent received over the measurement period divided by the value at beginning of the period:

$$IR = Y^{0-1} / CV^0$$

## Capital return

This is the change in value over the measurement period divided by the value at beginning of the period:

$$CR = [CV^1 - CV^0] / CV^0$$

## Total return

This is the sum of income return and capital return:

$$TR = [Y^{0-1} + CV^1 - CV^0] / CV^0$$

The following definitions describe the future.

## Required return

This is the return that needs to be produced by the investment to compensate the investor for the risks involved in holding the investment. It is also called the target rate or the hurdle rate of return. It is usually assessed as the sum of a risk free rate, such as the redemption yield on government fixed interest bonds or gilts, and a risk premium (see Baum (2009)). In this book we also used the term risk-adjusted discount rate or RADR for this concept.

## IRR or expected return

This is the expected return using estimated changes in rental value and yield. There are similarities in other investment markets: these include gross redemption yield, holding period return and IRR. The term 'equated yield' is sometimes used in property circles but serves only to confuse and should not be used.

This is the most accurate and complete description of historic return, defined as the discount rate needed to equate the PV of the future cash flows with the price. It is not a mean of annual total returns.

Given the total acquisition cost, the simplest level of analysis is to assess the initial yield. As defined above, this is the simple relationship or ratio of the first year's income and the total acquisition costs. It is known as or equivalent to the interest-only yield, the flat yield or the running yield in other investment classes.

Investors will also require an estimate of the expected return. This also helps the investment valuer to formulate his or her investment purchase advice. Continuing the example, we can estimate the expected resale price as follows:

| | |
|---|---|
| Implied rent in 2 years' time, as previously | £23,170 |
| PV £1 pa perp at 5% | 20 |
| Capital value | £463,400 |

The IRR (on an annual in arrears assumption) is found from the following cash flow:

| | |
|---|---|
| 0 | − £399,884 |
| 1 | + £9,500 |
| 2 + £463,400 + £9,500 = £472,900 | |

By trial and error, the IRR of this cash flow is 8.849%, delivered by (but not exactly equal to) a combination of the initial yield (2.376%) and the expected rental growth (7.6355%).

The majority of investors require yield calculations to be based on total acquisition cost. Solicitors' fees, surveyors' fees and stamp duty are the usual costs that need to be added to purchase price. In special cases it may be necessary to take account of immediate capital costs such as repairs, or in the case of a vacant building, refurbishment and letting expenses. In other cases it may be sensible to take account of rental apportionment and/or outstanding arrears of rents.

## Dealing with risk

A property investment is an exchange of capital today (current purchasing power) for future benefits. These future benefits may be in the form of income or capital growth, or a combination of both. As indicated earlier in this chapter, this requires the valuer to have some regard for the future and, as has so frequently been said before, the only certain thing about the future is its lack of certainty.

When an investor purchases future rights he or she has accepted 'risk'. When a valuer describes a property investment as being 'risky', he or she is implying some relative measure of uncertainty about the expected returns.

- The rents expected in the future may not be realised, so that for example expected rental growth will be less than anticipated.
- Increases in rent will not occur at the time expected or the property may become vacant and take some time to re-let.
- The capital value of the property on re-sale may not be realisable, may not increase with time or may fall with time.
- Costs associated with holding the property, such as repairs, may be unexpectedly high.

These property risks may be systematic or unsystematic (see Baum and Crosby (2008)). Tenant risk, sector risk, planning risk, and legal risk are unsystematic and, in a portfolio context, risk reduction can be achieved through diversification. Taxation, legislation and structural risks are more systematic and cannot be actively reduced through diversification. However, at an individual property level the investor is subject to specific or unsystematic risk, and in valuation the valuer is seeking to reflect the market's view of the risk relating to a specific property.

The more detailed the valuer's research into the financial stability of current tenants, the physical structure of the building, regional economics, and so on, the more able the valuer will be to express an expert opinion of open market value. The less certain the market is in respect of all these factors then the riskier the property will appear to be.

For this reason a valuer must reflect this greater risk in the valuation. The conventional income capitalisation approach adjusts for this through the use of a higher ARY based on market evidence of sales of higher risk properties (see Chapter 5). In growth-explicit DCF approaches, a common approach to dealing with risk is to increase the discount rate.

## The RADR

The use of RADR is still the most popular market approach used in growth-explicit DCF approaches. However, this leaves the valuer largely reliant upon a subjective or intuitive adjustment based on experience or market knowledge. How much should the discount rate be increased to allow for risk? Such an adjustment can be arbitrary. Unless market participants are familiar and comfortable with explicit valuation approaches, and exchange assumptions about rental growth rates and discount factors, the discount rate can have a subjective input.

Earlier, we estimated the discount rate by asking what target or hurdle rate the typical buyer of this property would use, locking this value to the assumed growth rate and the capitalisation rate.

Alternatively, we could estimate what hurdle rate we believe to be appropriate.

We have already seen from equation (2):

$$R = RFR + RP \tag{2}$$

Simplistically, the risk free rate (RFR) is the redemption yield on gilts for the matched life. To be accurate, the yield curve should be taken into account, meaning that the appropriate discount rate will be different for incomes of different maturity or tenor.

The required risk premium should be determined by the liquidity of the investment and by the sensitivity of the cash flow to shocks created by inaccurate forecasts or unforeseeable events.

The cash flow comprises the exit value and or any expected uplift at a rent review or lease renewal, and the sensitivity of the cash flow to economic shocks will be very important indeed. For those investors interested in the real cash flow, shocks to inflation may be important.

Default risk is also highly relevant. In addition, the risk premium will be affected by the extra illiquidity which affects all property, much more than listed bonds and equities.

## The sub-sector and property risk premium

As discussed in Baum (2009), it is common to include the provision of a series of risk premiums for sub-sectors of the property market, defined by sector (use type) and sub-sector, by region and by town. Where a sale or purchase is being assessed and the PV or net present value over purchase or sale price needs to be estimated, this process establishes a broad guide for estimating the risk premium which might be used in the discount rate. However, where an individual interest in property is being appraised, a further set of considerations needs to be taken into account.

Three main categories of premium can drive the specific risk premium in this particular system. These are: the sector (property type) or sub-sector premium; the town premium; and the property premium.

### The sector premium

Sector premiums are assessed by taking into account three factors. These are: the sensitivity of the cash flow to economic shocks, with particular reference to rental growth and depreciation; illiquidity; and other factors, including the impact on portfolio risk and the lease pattern. Are offices or retail properties more sensitive to an economic downturn?

### The town premium

The assessment of the town risk premium is based on an assessment of the riskiness of the economic structure of a town and its catchment area, together with a consideration of competing locations.

The range expands from a minimum town premium for diversified and liquid towns with healthy industries to maximum premiums for illiquid towns whose economies are concentrated in weak sectors.

Low liquidity scores are assigned to towns and sectors where it is considered relatively difficult to raise cash from a sale at short notice.

## The property premium

This section deals with the four components of the property premium, as listed below. The four components are:

a) the tenant risk class;

b) the lease risk class;

c) the building risk class; and

d) the location risk class.

Some of these factors will be specific to given sectors of the market (in-town retail, for example). The relative weighting of the factors can be assessed by multiple regression analysis, whereby (given a large sample of individual property investments) the current importance of these variables in explaining yield or risk premiums can be assessed and their future importance hypothesised.

The simple process is best illustrated by an example.

We are considering the purchase of either of two buildings. One (A) is a shop in Oxford; one (B) is a business park office near Reading.

Our estimate of the risk premium for a prime shop is 3% over the risk-free rate, currently 4.5%. Our estimate of the risk premium for a prime office is 4% over the risk-free rate, primarily to compensate for rental volatility and vacancy risk.

The tenant of property A is a FTSE 250 corporate; the tenant of property B is a partnership of solicitors. Additional premium: 0.5% for building B.

A is freehold; B is a leasehold for 116 years with 20% gearing (a ground rent of 20% of the passing sub-lease rent is payable). Additional premium for B: 1%.

The sub-lease for A has 18 years to run, with no breaks and upward-only rent reviews; B has 8 years to run, with a break clause which can be operated at year 3. Effectively, this removes the upward-only rent review. Additional premium for B: say 1%.

A is a simple, flexible building. B is at risk of depreciation, significant refurbishment expenses and vacancy. Additional premium for B: say 1.5%.

A is in a location in which competing supply is greatly restricted; B is surrounded by vacancy and planning restrictions in this area and for this sector are loose. Extra premium for B: 1%.

Table 12.1 summarises the cumulative effect of these individual adjustments.

Table 12.1: Building specific risk premia: an example

| Factor | Building A | Building B |
|---|---|---|
| Risk free rate | 4.50 | 4.50 |
| Base premium | 3.00 | 4.00 |
| Tenant | 0.00 | 0.50 |
| Tenure | 0.00 | 1.00 |
| Leases | 0.00 | 1.50 |
| Building | 0.00 | 1.00 |
| Location | 0.00 | 1.00 |
| Premium | 3.00 | 8.00 |
| Discount rate | 7.50 | 12.50 |

For a full worked example, see the case study described in Chapter 9 of Baum and Crosby (2008).

## Risk-adjusted cash flows: using sensitivity and simulation

The alternative to using RADR is to use statistical techniques to examine investment risk, and more specifically to model cash flow uncertainty. One such approach which is sometimes adopted is scenario analysis, whereby the valuer makes various estimates of the future – typically the best estimate, the most likely, and the worst – and estimates the probability of each occurring. The valuation would be an average of some sort, but adjusted by the range of the alternatives (a wide range suggests a risky investment and a lower valuation).

However, this approach ignores all the other possible outcomes. Consider the following example (taken from Byrne and Mackmin (1975)).

You are instructed by a banking organisation to prepare a valuation for mortgage purposes of a new owner-occupied office building. Having measured and surveyed the building and checked your findings with the architect's plans, you are satisfied that the building contains a total lettable area of 12,000 sq. m. It is your considered opinion, having regard to all the relevant factors, that the property would let at a figure between £250 and £300 per sq. metre and would sell as an investment on the basis of a 5-6.5% rate.

A preliminary valuation is prepared as follows:

| | |
|---|---|
| Area: | 12,000 sq m |
| Market rent: | £3,100,000 (approx £260 per sq m) |
| Yield: | 6% |
| | |
| Income | £3,100,000 |
| PV £1 pa in perp at 6% | 16.67 |
| Capital value | £51,677,000 |

However, this income capitalisation provided very little information. What is the expected cash flow? How risky is it? An explicit DCF valuation should involve some consideration of the variability of cash flows which is possible.

The ranges suggested for rents in the example are £250 to £300 per sq m and the yield range is 5 to 6.5%. Given these limits, and taking steps of, say, £5 in rent and 0.25% in yield, Table 12.2 shows the variations in final valuation obtained by altering these two variables within their respective ranges.

There are 77 possible 'outcomes' in Table 12.2. Which one is correct? Are any of them correct?

**Table 12.2** Rent in £ per sq m, Capital values in £m

| Rental | | | | Yield (%) | | | |
|--------|------|------|------|------|------|------|------|
| (£ sq m) | 5.0 | 5.25 | 5.50 | 5.75 | 6.0 | 6.25 | 6.50 |
| 250 | 60.00 | 57.14 | 54.55 | 52.17 | 50.00 | 48.00 | 46.15 |
| 255 | 61.20 | 58.29 | 55.64 | 53.21 | 51.00 | 48.96 | 47.08 |
| 260 | 62.40 | 59.43 | 56.73 | 54.26 | 52.00 | 49.92 | 48.00 |
| 265 | 63.60 | 60.57 | 57.82 | 55.30 | 53.00 | 59.88 | 48.92 |
| 270 | 64.80 | 61.71 | 58.91 | 56.44 | 54.00 | 51.84 | 49.85 |
| 275 | 66.00 | 62.86 | 60.00 | 57.39 | 55.00 | 52.80 | 50.77 |
| 280 | 67.20 | 64.00 | 61.09 | 58.43 | 56.00 | 53.76 | 51.69 |
| 285 | 68.40 | 65.14 | 62.18 | 59.48 | 57.00 | 54.72 | 52.62 |
| 290 | 69.60 | 66.29 | 63.27 | 60.52 | 58.00 | 55.68 | 53.54 |
| 295 | 70.80 | 67.43 | 64.36 | 61.57 | 59.00 | 56.04 | 54.46 |
| 300 | 72.00 | 68.57 | 65.45 | 62.61 | 60.00 | 57.60 | 55.38 |

Can the valuer justify his or her best assessment of £51.67 m – which is clearly only one of a much larger number of possible solutions – when this selection also implies the conscious rejection of, in this case, at least 76 other values? Is it possible to use the information at our disposal to arrive at a closer estimate of the likely value of this property?

Initially, a range of values between £46.15 m and £72.00 m may be noted. This range has a mean and a standard deviation. The latter is an accepted measure of risk. Can we use the range to define and adjust for the risk of the investment?

If an analysis of the data is made, it may be possible to determine the relative frequency of occurrence for the various rental levels between the minimum of £250 and maximum of £300.

Let us suppose that for this example such an analysis is possible for 50 comparable transactions: the results can then be tabulated, as in Table 12.3. This gives a good indication of the probability of

**Table 12.3**

| Rental (£ sq m) | Frequency | % Occurrence | Probability |
|---|---|---|---|
| 250 | 1 | 2.0 | 0.02 |
| 255 | 2 | 4.0 | 0.04 |
| 260 | 5 | 10.0 | 0.10 |
| 265 | 6 | 12.0 | 0.12 |
| 270 | 7 | 14.0 | 0.14 |
| 275 | 9 | 18.0 | 0.18 |
| 280 | 7 | 14.0 | 0.14 |
| 285 | 6 | 12.0 | 0.12 |
| 290 | 5 | 10.0 | 1.10 |
| 295 | 1 | 2.0 | 0.02 |
| 300 | 1 | 2.0 | 0.02 |
| Total | 50 | 100.0 | 1.00 |

occurrence of the various possible rentals presupposing, of course, that the 50 transactions are truly comparable. If insufficient data is available it may be necessary to use other methods, as described below.

Each rental may now be 'weighted' by multiplying it by its probability of occurrence and summated to give one overall expected rental value, each element being included in proportion to the probability of its occurrence. All possible rental values will then have been built into the result. None are discarded at this stage, but their importance is now related to the known frequency of occurrence of each rental level. Since the distribution of probabilities in Table 12.2 is almost symmetric, then the expected rental will be in the centre of the distribution. In this case it is £275.

The expected rental obtained here is specific to the distribution shown in Table 12.3; other shapes of probability distribution can occur, and when this happens the expected value will be different. An analysis of transactions, for example, might show a different frequency and probability pattern, as in Table 12.4.

(It is not unreasonable to argue that in each of these distributions the 'modal value' (that rental having the largest observed frequency of occurrence) could be taken as representative, since it is the most likely value.)

In Table 12.4, the relative frequency of rents shows that some rentals are quite probable, £265 for example. The use of the expected value, the weighted mean, reflects the possibility that these other results might occur. Whereas the central value appears to be £275, the expected rental value in this case is £265, showing that, in spite of the evidence that 40% of observed rentals are at £260, 46% are above £260, and 14% below.

**Table 12.4**

| Rental (£ sq m) | Frequency | % Occurrence | Probability |
|---|---|---|---|
| 250 | 2 | 4.0 | 0.04 |
| 255 | 5 | 10.0 | 0.10 |
| 260 | 20 | 40.0 | 0.40 |
| 265 | 10 | 20.0 | 0.20 |
| 270 | 4 | 8.0 | 0.08 |
| 275 | 2 | 4.0 | 0.04 |
| 280 | 3 | 6.0 | 0.06 |
| 285 | 2 | 4.0 | 0.04 |
| 290 | 1 | 2.0 | 0.02 |
| 295 | 1 | 2.0 | 0.02 |
| 300 | 0 | 0.0 | 0.00 |
| Total | 50 | 100.0 | 1.00 |

**Table 12.5** (values in £m)

| | Rental (£/sq m) | |
|---|---|---|
| Yield (%) | £275 | £265 |
| 5.00 | 66.00 | 63.60 |
| 5.25 | 62.86 | 60.57 |
| 5.50 | 60.00 | 57.82 |
| 5.75 | 57.39 | 55.30 |
| 6.00 | 55.00 | 53.00 |
| 6.25 | 52.80 | 50.88 |
| 6.50 | 50.77 | 48.92 |

The different results from these two sample rents are compared in Table 12.5 for the seven possible yields used before. This range of values may be acceptable if the number of alternatives remains relatively small. The range of possible alternatives may be very large, however and, more importantly, it may be possible to determine how likely they are to occur by means of an analysis of observed frequencies.

Tables 12.2 and 12.3 show two different distributions of values. The first is close to a normal distribution; the second is clearly non-normal or skewed (negatively, or to the low side). Non-normal distributions are very common in property and make life more difficult, because the central value may not be the weighted average or expected value.

**Table 12.6**

| Yield (%) | Probability |
| --- | --- |
| 5.00 | 0.02 |
| 5.25 | 0.03 |
| 5.50 | 0.05 |
| 5.75 | 0.15 |
| 6.00 | 0.45 |
| 6.25 | 0.20 |
| 6.50 | 0.10 |
| Total | 1.00 |

There are two possible ways of dealing with this problem:

- Use a computer to calculate and display the results for all possible alternatives. (Such an exercise is called 'a simulation'.)
- Make use of available experience to determine subjective probabilities for the occurrence of 'likely' values for these variables. These probabilities can then be built into the consideration of alternatives.

In the example above, Tables 12.3 and 12.4, the various values were all suggested as equally possible. Clearly, the valuer should be able to say from his or her knowledge that all are not equally possible, but that some are most unlikely and, more importantly, that some are very likely. This view is based upon the considered opinion of the valuer.

There are standard and easily-learned rules to enable such considered opinions to be converted to subjective probabilities recognised to be as valid as the objective assessments derived from long-run frequencies, as shown in Tables 12.3 and 12.4. A complete probability distribution can be built up for any variable using these methods.

As has been seen earlier, yields are just as likely to vary as rentals in this example. After consideration, the following subjective probabilities have been placed against the possible yields (see Table 12.6).

The distribution of probabilities is such that only a few yields are considered likely to occur. From this distribution, the expected value for the yield may be obtained by weighting each yield by its probability: in this case the value is 5.995% (6.00%).

The implication of the subjective selection of a high probability of occurrence for a particular yield is that it is considered relatively risk free. It is also possible, therefore, to use such assessments as indicators of individuals' attitudes to risk.

The PV factor may then be calculated, and the capital value arrived at in the usual way.

Distribution 1 (Table 12.3)

| | |
|---|---|
| Estimated yield | 6% |
| Estimated full rental value 12,000 × £275 | £33,000 |
| PV £1 pa perp at 6% | 16.67 |
| | £550,110 |

Distribution 2 (Table 13.3)

| | |
|---|---|
| Estimated yield | 6% |
| Estimated full rental value £12,000 × £265 | £31,800 |
| PV £1 pa at 6% | 16.67 |
| | £530,106 |

The method outlined is part of a more scientific approach to valuation (see Baum and Crosby (2008) for a more detailed analysis of this type).

These valuations may be compared with the original best estimate or preliminary valuation. Any solution derived in this way must be understood to be an estimate. The use of a statistical analysis of this type can only produce estimates. However, and this is more important, it produces a more consistent approach to uncertain situations, highlighting the stages in the appraisal process and pointing to any inconsistencies requiring correction or modification.

Every input, variable or otherwise, is dependent upon the strength of the valuer's evidence, the valuer's rental and value analysis, and the valuer's understanding and assessment of current market conditions. The approach differs in that the valuer is required to consider ranges of uncertain variables much more carefully, any single figure arrived at being recognised as an estimate based on a proper analysis of the market.

Using a method such as this, a valuer may quite reasonably derive a series of results for any valuation. In that case, although there are no reasons why a range of valuations may not be very helpful, great care must be taken in presenting such findings to the client who expects a single valuation result. The complete findings should be incorporated into valuation reports as appendices.

Naturally, the method should not be applied automatically. It requires a clear understanding of the statistical methodology and its implications; and it could also be inappropriate in some situations.

Many valuers see little need for the explicit use of probability in their valuations. In this book it has been indicated that on many occasions there is a lack of certainty. One cannot be certain what the rent of a vacant building really is until it is actually let, one cannot be certain what rent will be achieved on review, one cannot be certain of the capitalisation rates to be used, and one cannot be certain about future costs on repairs and refurbishment. There is a risk; why not reflect it?

## Risk-adjusted valuations

In the market, the use of simulations has been largely restricted to development appraisal work. Sophisticated approaches are now in place in many property organisations; statistical packages such as Crystal Ball and @RISK are available off the shelf for model-building; and property cash flow simulations are also available.

Valuers are naturally resistant to such approaches. An argument for adhering to market capitalisation approaches rests with the well-established view that accuracy in assessment of the key variables at today's date is sufficiently problematic without adding the difficulty of estimating future rental levels and capitalisation or discount rates. The further the valuer ventures away from the market, the greater is the probability that one is adding human error to the already high levels of risk and uncertainty that attach to property.

Nonetheless, valuers are regularly encouraged to embrace this type of technology. Baum and Crosby (2008) explore the use of sensitivity analysis. This they regard as 'a somewhat rudimentary risk analysis technique which helps investors to arrive at a decision but fails to identify the chances of the possible variations becoming fact'. Recognising the popularity of the RADR approach, they also explore and recommend the use of 'Certainty Equivalent Cash Flow Models', in particular the use of the hybrid 'Sliced Income Approach'.

These models make use of normal distribution theory which holds that some 68% of all values in a normal distribution will be within + 1 standard deviation of the mean; 95% within + 2; and 99% within + 3 standard deviations. From this a certain equivalent cash flow can be constructed. Both Baum and Crosby (2008) and Dubben and Sayce (1991) apply this approach to property investment cash flows.

Given the best estimate of rent minus one standard deviation, the best estimate of growth minus one standard deviation, the best estimate of yield plus one standard deviation, and a risk free discount rate, a certainty equivalent (CE) cash flow can be constructed.

The following is based on Dubben and Sayce's example of a property let at £10,000 for two years with an ERV of £15,000. Given this data, the cash flow could be considered as:

Current rent: £10,000 pa for 2 years.

ERV: £15,000 pa minus 1 standard deviation (£1,000)
CE = £14,000
Rental growth: 5% minus 1 standard deviation (1%)
CE = 4%
Capitalisation rate: 6% plus 1 standard deviation (0.5%)
CE = 6.5%

It will be noted that this is a risk-averse method, as negative scenarios have been adopted for each variable. This, it is suggested, provides a CE value.

There are doubts about the usefulness of this technique when weighed against market experience. What (for example) is the certainty equivalent of a fixed contracted cash flow paid by a FTSE 250 tenant? The sliced income approach is a more logical model for the typical property investment as it accepts the fact that the current rent payable under a lease is relatively risk free. The riskiness is likely to be a function of a tenant's ability to pay the rent, which can be judged reasonably accurately given a thorough assessment of the tenant's credit rating. This method allows the income to be split between the current rent which is certain, and the future rent, which is less certain. Given this, it is possible to assess the value of the certain rent at the risk free rate.

There is little doubt that if valuers use discounted cash flow, or growth-explicit, approaches to valuation, RADR approaches are clearly preferred in practice – but this is not to say that risk adjusted cash flows are not helpful in property investment analysis, and more examples of applications will emerge in time.

## Risk-adjusted IRRs

A number of writers believe that investment advice needs to be based on a more rigorous cash flow approach. This process can be very simple or very sophisticated, but in either form provides the analyst with the opportunity to test the sensitivity of the investment to a variety of variables, including rent review costs and lease renewal costs; voids and refurbishment costs; other non-recoverable service or repair costs; sale costs; income and capital gains tax; depreciation; exit yields and residual values. Such modelling also allows the valuer to test the effect of different rental growth factors, either on an average long-term basis or by adopting different rates of growth over different time periods.

It has already been indicated that the complexities of certain geared leaseholds and development funding schemes can only effectively be handled in this way, and sometimes the value of an investment can only be revealed by such a simulation approach (see Baum and Crosby (2008), Chapter 9). The major advantage of simulation is that the probability of achieving the stated yields or return can be estimated. Consider Example 12.1.

### Example 12.1

Your clients are contemplating the acquisition of a property investment. The property is fully let, producing a net cash flow from contracted rents of £1,000 per annum for the next two years. These rents are considered to be certain. In two years' time the rents are due for review. You have considered the likely level of rents in two years' time and expect the rent roll to be in the region of £3,000. You have prepared the following table of cash flows, together with a probability measure for the capital reversions in two years' time. The asking price plus acquisition cost gives a total purchase price of £32,640. Advise your client.

| End year | Net sum | Probability | Expectation |
|----------|---------|-------------|-------------|
| 1 | £1,000 | 1 | £1,000 |
| 2 | £1,000 | 1 | £1,000 |
| 2 | £30,000 | 0.20 | £6,000 |

| End year | Net sum | Probability | Expectation |
|----------|---------|-------------|-------------|
| 2 | £37,500 | 0.60 | £22,500 |
| 2 | £40,000 | 0.20 | £8,000 |

The probability-adjusted cash flow in year 2 is the sum of these values, i.e. a rent of £1,000 and a weighted average capital receipt of £36,500. The internal rate of return is virtually 8%. If the client's target rate is 7% an investment is worthwhile.

# Regression analysis

Regression analysis is a standard tool used in investment analysis. In its simplest form, the statistical technique of regression analysis enables the analyst to predict the value of one variable from the known value of another. Valuation would be an extremely simple science if, for example, the valuer could predict the sale price of a property from its total floor area. Assuming the simplest possible case, this might indicate that a linear relationship between value and size existed which, if plotted on a graph, would produce a straight line, the slope of which would be the variable factor, b. Where a is a constant, the standard formula for a linear relationship is:

$$y = a + bx$$

In the simple case selected above, y represents house value and x represents total floor area. For example, it might be noted that in a given locality the price of houses was always equal to £5,000 plus the floor area multiplied by 5. Then:

$$y = £5,000 + 5x$$

However, it is a normal presumption that value is a function of a number of variables. Multiple linear regression enables the analyst to bring into play as many variables that may be considered to be likely to affect the value or likely selling price of the subject property, such as parking facilities, outlook, location, size, specific facilities such as central heating and garage space and other factors such as the age and condition of the building.

Regression packages developed for property provide a ready-made analysis of the relevant inputs, which will typically consist of a list of property characteristics for the type of property under analysis (the independent variables), together with details of the actual sale prices or rentals achieved (the dependent variable). The model correlates each feature with the known factor, sale price, selects those features with the highest correlation, produces a regression equation and estimates the sale prices on that basis. The computer then calculates the difference between actual price and the estimated price, and then proceeds to select from the remaining features the next highest correlation and proceeds until all the features have been used.

The end result is an equation which can, with care, be used to estimate sale price, or whatever other factor is required in respect of another property. As more data becomes available, this is added to the existing store and the program re-run to check for any significant changes in preferences by purchasers.

Any predicted figure must not be regarded as an absolute, and the valuer requires some indication of accuracy or acceptability. The statistical measure of accuracy used is generally R (the correlation coefficient) and $R^2$ (the coefficient of determination).

When the data produces a value for R as close as possible to 1 this would imply that the variation of the dependent variable (sale price, rental value) is explained fully by the independent variable(s). If R falls below 0.9 then only about 80% of the variation of the dependent variable is explained by the independent variable(s) and the smaller R becomes, the less meaningful is the whole analysis.

The use of multiple regression analysis has been primarily limited to mass valuation problems, particularly for land-tax purposes and residential property. The majority of reported examples concern house prices, but it is suggested that regression and multiple regression could be used for other applications, such as:

- Testing the relationship between size and price or rent (for example, the rate at which land price per hectare decreases with the increase in the size of the holding being sold, or the extent to which rents per square metre for office or industrial space decrease with the size of the letting).
- Time/value trend analysis.
- Determining the rental value of all types of premises (as rent is a function of size, location, facilities, running costs, consumer income).
- Predicting gross trading income from licensed premises, theatres, restaurants, etc from the number of persons using the premises.
- Predicting petrol throughput for service stations based on traffic counts and other variables.
- Estimating repair expenses based on property maintenance records.

Clearly, valuations and analyses of the type outlined in this chapter are only likely to be undertaken by the larger national valuation practices, government and institutional investors. For these reasons the description of the alternative techniques used has been kept to a minimum to give the average reader and student valuer a general idea of the developing techniques. Readers requiring a more detailed approach to these techniques are referred to the recommended reading at the end of the book.

We began by suggesting that valuers should keep their minds open to new techniques. A major development in appraisal techniques in recent times has been the use of statistics. Norman Benedict aptly summarised our own feelings in an article published in the Appraisal Journal as long ago as October 1972:

> Statistical analysis can significantly broaden the role of the appraiser and substantially increase his effectiveness, providing him with the tools to attain greater sophistication and expertise in the areas of marketability, feasibility and investment analysis.
>
> Correspondingly, the appraiser who clings to yesterday's tools to meet tomorrow's challenges will become progressively less effective and less involved in his work, while the appraiser who seeks constantly to acquire new skills will develop both personally and professionally. In summation, then, statistics represents a golden opportunity for an appraiser to experience both personal and professional growth.

This statement is as apt today as it was in 1972. Its pertinence is obvious to those valuers developing or using statistical or econometric models, and who are willing to tackle the critical area of property market forecasting. There are many opportunities for practitioners to develop the proper use of these techniques in responding to their clients' needs. After all, an opinion of value is easily challenged; it will only be defensible if it is supported by best practice analysis of good available evidence.

## Questions

Define:

a) IRR.
b) Initial yield.
c) Equivalent yield.
d) Required return.

A property has just been sold for £1 million freehold. It is currently producing a rent of £60,000 pa net. The full rental value is £110,000 pa. The lease will be renewed in 4 years time.

Calculate the following:

a) The total acquisition costs with fees and stamp duties at 5.67%.
b) The current return based on purchase price plus acquisition costs.
c) The expected return after the lease is renewed.
d) The capitalisation rate(s) used by the valuer.
e) The IRR assuming rental growth at 5% per annum and a re-sale in four years time; assuming no change in the capitalisation rate.
f) The growth in rental value needed over the next four years:

    i) in pounds.
    ii) as a rate of growth per annum in order to achieve a required return (IRR) of 12%.

# Spreadsheet User

The Excel spreadsheet facility now available to most valuers provides a flexible tool for addressing typical tasks which are required on a repeat basis by clients. One such example would be the use of Excel for the purpose of calculating worth to an investor client having agreed the inputs, namely:

- holding period
- rental growth
- client's target rate
- factual information as per current lease terms, rents, rent reviews, outgoings, etc.
- variables such as non-recoverable management costs, UBR, service costs, lease renewal fees, rent review fees and the like.
- capitalisation rates (ARYs), particularly for the assessment of terminal value at the end of the holding period; or as a basis for assessing site value at the end of the lease.

Simulation will convince most readers that the key variables in a worth calculation are those relating to rental growth and the target rate. Rental growth should be based on reasoned argument having regard to macro and micro economic conditions.

The illustration here is simply an illustration, but creating an Excel spreadsheet for investment worth/value calculations is our final challenge for the reader/learner who has got this far. Please improve on our efforts and move to a quarterly in advance format.

# INVESTMENT WORTH CALCULATIONS

## DATA

| | |
|---|---|
| CURRENT RENT | £100,000 |
| MARKET RENTAL VALUE | £150,000 |
| RENTAL GROWTH RATE | 3.50% |
| ALL RISKS YIELD AT END OF HOLDING PERIOD | 7% |
| INVESTORS TARGET RATE | 10% |
| MANAGEMENT FEES | £5,000 |
| RENT REVIEW FEES | 5% |
| LEASE RENEWAL FEES | 7% |
| RENT/LEASE RENEWALDATE | 2009 2014 |
| MANAGEMENT FEES INFLATION INCREASE ADJUSTMENT | 3% |
| DATE OF ANALYSIS | 2006 |

## ANNUAL IN ARREARS ASSUMPTIONS

| Period | MRV + Growth | Rent | Management fee | RR/LRFees | Net Income | PV at H9 | PV M * K |
|---|---|---|---|---|---|---|---|
| 0 | £150,000 | | £5,000 | | | | |
| 1 | £155,250 | £100,000 | 5150 | | £94,850 | 0.909090909 | £86,227 |
| 2 | £160,684 | £100,000 | 5305 | | £94,696 | 0.826446281 | £78,261 |
| 3 | £166,308 | £100,000 | 5464 | | £94,536 | 0.751314801 | £71,027 |
| 4 | £172,128 | £166,308 | 5628 | 8,315 | £152,365 | 0.683013455 | £104,067 |
| 5 | £178,153 | £166,308 | 5796 | | £160,512 | 0.620921323 | £99,665 |
| 6 | £184,388 | £166,308 | 5970 | | £160,338 | 0.56447393 | £90,506 |
| 7 | £190,842 | £166,308 | 6149 | | £160,159 | 0.513158118 | £82,187 |
| 8 | £197,521 | £197,521 | 6334 | | £159,974 | 0.46650738 | £74,629 |
| 9 | £204,435 | £197,521 | 6524 | 13,826 | £177,171 | 0.424097618 | £75,138 |
| 10 | £211,590 | £197,521 | 6720 | | £190,801 | 0.385543289 | £73,562 |
| Sale price | | | | | £2,985,789 | 0.385543289 | £1,151,151 |
| INVESTMENT WORTH | | | | | | | £1,986,420 |

The sale price in year 10 is based on a valuation at that point in time. In this case this is £197,521 for 3 years plus £211,590 in perpetuity deferred 3 years all at 7%.

## Dual rate

As noted in Chapter 2, dual rate methodology is no longer considered to be appropriate for the valuation of leasehold interests in property. In the recent publication by the Royal Institution of Chartered Surveyors (RICS) *Valuation Calculations: 101 Worked Examples* (2010) by Ollie Saunders, a Valuation Partner in Drivers Jonas Deloitte, single rate and discounted cash flow appear to be the preferred methods for leasehold valuations. This Appendix is included for the benefit of educators who find the subject matter to be of 'academic' interest, and for those valuers who still use the method.

One view of the present value of £1 pa is that it can be considered to be replacing capital at the same rate as the remunerative or investor's rate of return on capital. As such, the sinking fund is notional and ensures that the correct value is assigned to the investment. Single rate is based on the principle of the rate of return being the internal rate of return (IRR). This provides for the return on money invested to be considered as a return on the amount of capital owed, or outstanding, from year to year. This principle has already been illustrated in relation to annuities and mortgages. Dual rate is based on the principle of the rate of return remaining constant from year to year, whilst the recovery, or redemption of capital occurs through the regular reinvestment of part of the income in a separate sinking fund which may be accumulating at the same rate as the rate of return on capital or, more typically, at a lower guaranteed rate of interest.

Despite the fact that investors rarely, if ever, provide for replacement of capital in this way; a method developed from the start of the twentieth century for the valuation of limited term property investments based upon this sinking fund concept.

The technique is only used by some valuers to value wasting assets such as leasehold interests, it is not the basis of assessing the value or price of other time limited investments, such as dated government stocks, and the application of the method with a specific low sinking fund rate has been, for the most part, restricted in use to the United Kingdom. A freehold interest in property could be likened to saving in a building society (the principal is retained in the ownership of the investor) and all income represents a return on capital employed. A leasehold interest is a terminating or wasting asset and comes to an end after a given number of years. The sum originally invested is spent in return for an income for a given number of years. At the end of that time nothing remains. It is just like a mortgage.

The valuer is normally under instruction to assess the market value of an interest in property. A principle of market valuation is comparison, and the responsibility of the valuer is to analyse the market so as to be able to assess the market value of other property by comparison. The only tool of analysis and comparison in the investment market is the IRR (single rate) as there can only ever be one IRR for an investment under normal circumstances. To analyse the sale price of a time limited interest in property on a dual rate basis requires the analyst to assume the sinking fund rate (known as the accumulative rate). That assumption then determines the rate of return or remunerative rate. It is thus possible for different analysts to arrive at different views as to the return from a property, creating problems of market comparison. The valuer has a further responsibility which is to ensure that his or her methods of market valuation mirror the behaviour of the buyers and sellers in a particular market. As there is no evidence of investors in property reinvesting in sinking funds, there would appear to be no justification for valuers to make such assumptions in their valuation of property.

These arguments are ignored by those valuers whose education and training has taught them that dual rate must be used. The rest of this Appendix looks at some of the dual rate problems, and as normally described by UK valuers as years' purchase (YP) dual rate we have stuck with this terminology. Readers are referred to Baum and Crosby (2008) *Property Investment Appraisal*, Estates Gazette for a complete treatise on dual rate methodology.

### YP dual rate (PV £1 pa dual rate)

In order to compare the return from an investment in a leasehold interest with an investment in a freehold interest, it is argued that the original outlay must be returned at the end of the lease, so that a similar and equal income flow may be acquired. This process may continue into perpetuity so that a perpetual income may be enjoyed, and the return becomes comparable with that receivable from a freehold investment.

There would seem to be no problem at first glance as present value (PV) £1 pa includes a sinking fund. However, two problems are encountered if an actual sinking fund is to be arranged. Single rate assumes that the sinking fund accumulates at the same rate as the yield from the investment, the rate of interest at which a sinking fund will accumulate does not necessarily relate to the yield given by the investment itself. Two rates may thus be needed. Second, the sinking fund must replace the initial capital outlay and so the possible effect of tax on that part of the income that represents capital replacement cannot be ignored; nor can the effect of tax on sinking fund accumulations.

As already stated, there is no reason to suppose that the 'accumulative rate' will equate with the yield from the investment or the 'remunerative rate'. Where it does not, the YP figure will be 'dual rate'.

The only formula for a dual rate YP, catering for the difference in accumulative and remunerative rates, is:

$$\frac{1}{i+sf}$$

The following formula can only be used in the case of a single rate PV £1 pa:

$$\frac{1-PV}{i}$$

The following formula can be used for single or dual rates:

$$\frac{1}{i+sf}$$

---

**Example A.1**

Value a limited income of £1,500 pa for six years where your client requires a yield of 11% and the best safe accumulative rate is 3%. Show how a return on and a return of capital are received.

| | |
|---|---|
| Income | £1,500.00 |
| PV £1 pa for 6 years at 11% and 3% | 3.7793 |
| Capital value | £5,668.95 |

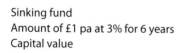

| Return on capital | Return of capital |
|---|---|
| = 0.11 x £5,669 | = £1,500 – £623.59 |
| = £623.59 | = £876.41 |

| | |
|---|---|
| Sinking fund | £876.41 |
| Amount of £1 pa at 3% for 6 years | 6.4684 |
| Capital value | £5,668.97 |

## Adjustment for tax

The use of dual rate assumes that sinking funds are taken out in practice. Such sinking funds are designed to replace the initial outlay on an investment. As the sinking fund has to perform this function without question, it is assumed to accumulate at a net of tax rate.

If an investor pays tax at a rate of say 25%, and the rate of interest earned by the sinking fund is subject to this rate of tax, then a gross accumulative rate will be reduced to a net accumulative rate.

For example, where an annual sinking fund (ASF) is £150 pa and the gross accumulative rate is 5% with tax at 25%, then after one year the interest earned is:

£150 × 5% = £7.50 but tax at 25% reduces this to (0.75 × £7.50) = £5.625.

£5.625 is only 3.75% of £150 so the gross rate of 5% has been reduced to a net accumulative rate of 3.75%.

Such a calculation is easily accomplished by applying a tax adjustment factor of $1 - t$ (where $t$ = the rate of tax expressed as a decimal) to the original gross rate.

If the required sinking fund instalment is calculated on gross instead of net rates of interest, it will simply be inadequate whenever interest on the sinking fund is taxed.

---

**Example A.2**

£1,000 must be replaced within 10 years. The accumulative rate is 6%, calculate the ASF.

| | |
|---|---:|
| Sum required | £1,000 |
| ASF to replace £1 in 10 years at 6% | 0.075868 |
| | £75.87 |

However, the sinking fund is taxed at 25%. It will therefore actually accumulate at 6% multiplied by the tax adjustment factor of $(1 - t)$:

| | |
|---|---|
| = | 6% (1 – t) |
| = | 6% (0.75) |
| = | 4.5% |

| | |
|---|---:|
| ASF | £75.87 |
| A £1 pa for 10 years at 4.5% | 12.29 |
| Capital replaced | £932.44 |

The sinking fund is insufficient to replace the initial outlay of £1,000, due to the effect of tax on the sinking fund accumulation. The sinking fund must be calculated in the light of the tax rate:

| | |
|---|---:|
| Sum required | £1,000 |
| ASF to replace £1 in 10 years at 4.5% | 0.08137 |
| ASF | £81.37 |
| A £1 pa for 10 years at 4.5% | 12.29 |
| Capital replaced | £1,000 |

The sinking fund has this time been correctly calculated to accumulate after the effect of 25% tax on the interest accumulating in the sinking fund. Accumulative rates must be net of tax to compensate for this first effect that tax has on the accumulation of the sinking fund.

Tax also affects income from property. Rates of return from most investments are quoted gross of tax, because individual tax rates vary and net of tax comparisons may, as a result, be meaningless. The remunerative rate $i\%$ in a dual rate YP is therefore a gross rate of interest.

However, a sinking fund is tied to the principle that it must actually replace the initial capital outlay so that a comparable investment may be purchased. The effect of tax on the income cannot therefore be ignored.

### Example A.3a

Value a profit rent of £2,000 pa receivable for 10 years using a remunerative rate of 10% gross and an accumulative rate of 3% net. The investor pays tax at 25p in the £ on all property income. Show how the calculation is affected.

| | |
|---|---|
| Ignoring tax on income: | £2,000 |
| PV £1 pa for 10 years at 10% and 3% | 5.341 |
| Capital value | £10,682 |

But income is taxed at 25p in the £1:

Net income $= £2,000 \times (1 - t)$
$= £2,000\,(1 - 0.25)$
$= £2,000\,(0.75)$
$= £1,500$

From this net income a net remunerative rate of 7.5% [10% $(1 - t)$] and a sinking fund to replace the initial capital outlay must be found.

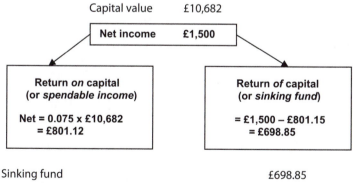

| Capital value | £10,682 |
|---|---|
| Net income | £1,500 |

**Return *on* capital (or *spendable income*)**

Net $= 0.075 \times £10,682$
$= £801.12$

**Return *of* capital (or *sinking fund*)**

$= £1,500 - £801.15$
$= £698.85$

| | |
|---|---|
| Sinking fund | £698.85 |
| A £1 pa for 10 years at 3% | 11.4639 |
| | £8,011.55 |

The sum calculated above fails to replace the initial outlay of £10,682.

Why?

Answer: As the income is reduced by 25%, both the spendable income and the sinking fund contributions must be reduced by 25%. The spendable income then becomes a net spendable income representing a net return on capital and still conforms to the investor's requirements. However, the net sinking fund must replace £10,682; the fact that the gross sinking fund would notionally replace the initial outlay is no comfort for the investor who is left several thousand pounds short.

It must therefore be ensured that the net sinking fund still replaces the initial outlay.

Thus:
Gross sinking fund $\times (1 - t) =$ net sinking fund

$$\text{Gross sinking fund} = netSF \times \frac{1}{(1 - t)}$$

If, having calculated the desired amount that should remain as a net sinking fund, this amount is multiplied by the 'grossing-up' factor of $1/(1 - t)$ then the required amount of net sinking fund will remain available after tax.

Clearly therefore more income must be set aside as gross sinking fund. This means that the amount of income remaining as spendable income will be reduced, and the investor's requirements of a 10% return will not be fulfilled. A new valuation is therefore required, reducing the price paid so that a grossed-up sinking fund may be provided and a 10% return (gross) can still be attained.

A new PV £1 pa figure must be calculated using a 10% remunerative rate, a 3% net accumulative rate, and a grossing-up factor applied to the SF of $1/(1 - t)$.

The dual rate formula adjusted for tax is therefore:

$$\frac{1}{i + (SF \times \frac{1}{(1 - t)})}$$

In this case this becomes:

$$\frac{1}{0.10 + [0.08723 \times \frac{1}{1 - 0.25}]}$$

$$= \frac{1}{0.10 + [0.08723 \times \frac{1}{0.75}]}$$

$$= \frac{1}{0.10 + 0.11630}$$

$$= \frac{1}{0.2163}$$

$$= 4.6232$$

**Example A.3b**

Revaluation of Example A.3a using the dual rate YP formula:

| | |
|---|---|
| Income: | £2,000 |
| YP for 10 years at 10% and 3% adj tax at 25% | 4.6232 |
| Capital value | £9,246 |

| Proof: | Capital value | £9,246 |
|---|---|---|
| | Gross income | £2,000 |
| | Net income | £1,500 |

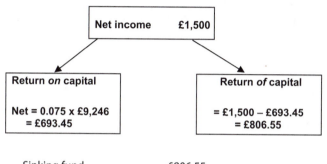

| | |
|---|---|
| Sinking fund | £806.55 |
| A £1 pa for 10 years at 3% | 11.4639 |
| Capital Value | £9,246 |

The result of using a tax-adjusted dual rate has been to reduce the capital value from £10,682 to £9,246. This has enabled the investor to gross-up the sinking fund to compensate for the effect of income tax, and leave enough income after tax to provide a 10% gross and a 7.5% net return on capital.

On a single rate basis, a sale of an investment income of £2,000 a year for £9,246 is simply analysed to assess the IRR (see Chapter 3). In this instance the IRR is 17.21%, or on a net of tax basis with tax at 25%, which reduces the £2,000 to £1,500, the IRR is 9.93%. In practice, as valuation is an imprecise science, valuers would tend to round the comparable information to 10% net. This single rate basis translated to normal investment terminology of IRR instantly communicates with investors. Terms such as dual rate adjusted for tax have no meaning to investors.

## Dual rate review and issues

This section has been taken from Chapter 7 of the fifth edition of this book. A few minor changes have been made to reflect changes following the credit crunch.

The reasons generally given for valuing leaseholds on the dual rate basis for recoupment (recovery) of capital by reinvestment in a sinking fund at a low safe rate are as follows:

1. That an investor requires a return on initial outlay throughout the lease term at the remunerative rate.

2. That all capital must be recovered by the end of the lease term and that to be certain of doing this reinvestment in a sinking fund accumulating at a low, safe, net of tax rate must be undertaken.
3. That by so doing the investor has equated the finite investment with an infinite (freehold) investment in the same property base, i.e. the spendable income is perpetuated.

### Example A.4

Estimate the value of a four-year unexpired lease.
The profit rent is £100 per annum net.

| | |
|---|---:|
| Net of outgoings profit rent | £100 |
| YP £1 pa for 4 years at 10% and 3% adj. tax at 40% | 2.00 |
| | £200 |

Proof:

ASF to replace £1 at 3% in 4 years = 0.239

ASF to replace £200      = 200 × 0.239

     = £47.80

Grossed-up to allow for tax on original income of £100

$$= £47.80 \times \frac{1}{1-t}$$

$$= £47.80 \times \frac{1}{1-0.40}$$

$$= £47.80 \times \frac{1}{1-0.60}$$

$$= £79.66$$

Grossed-up sinking fund = say £80
Therefore:

Spendable income      = £100 – £80

     = £20

     = a 10% return on £200

In theory the £200 recovered over four years can be reinvested at 10% to produce a 10% return in perpetuity.

However, note:
1. Capitalisation on a single rate basis can be considered to give a return on initial outlay plus a sinking fund. Additionally it can be shown to give a return on capital at risk.
2. It can be shown that when a single rate years' purchase is used, the investor will, at the implied sinking fund rate, recover all capital by the end of the term.
3. The investment may have been perpetuated and can be shown to be perpetuated on a single rate and dual rate basis. However, a leasehold interest cannot be equated with a freehold interest if one merely recovers historic cost during periods of inflation.

Within conventional dual rate valuations two schools of thought are apparent. Initially both accept that the risk or remunerative rate for leasehold should be a minimum of 1 to 2% above the appropriate freehold rate. This is to allow for the additional problems of reinvesting in a sinking fund, reinvesting the accumulated sinking fund at the end of the lease term, and for the additional risks inherent in a leasehold investment relating to restrictions imposed by the lease terms and possible dilapidations. This philosophy would have been sound in the late nineteenth and early twentieth century when leases were for long terms of 99 years or more. To assume some constant magical relationship today is inconsistent with the philosophy of market valuations, unless market transactions justify that rate.

Both schools assume that reinvestment must be at a low, net of tax rate and that the sinking fund premium or instalment must be met out of taxed income. However, the schools seem to differ on the nature of the sinking fund.

The original theory was that capital replacement, in order to be guaranteed, must be in effect 'assured'. Hence reinvestment was assumed to be made in a sinking fund policy with an assurance company. Such a policy is a legal contract. Such a policy is unaffected by any changes in tax rates, as such changes will only affect the investor's investment income. (If a policy is taken out and a resale occurs part way through the term, the insured must either continue to pay the annual premium or accept a paid-up policy. In the latter case there would be a shortfall in capital recovery.) The total sum assured will be paid out on the due date if all premiums have been met. The rate of accumulation is assumed to be between 2.5% and 4%; however by 2010 a rate of 0.5% to 1.0%, net of tax is more realistic as the base rate stays at 0.5%.

The second school argues in favour of reinvestment at safe rates. These will not necessarily be guaranteed but could conceivably include regular savings in a post office account, bank deposit account, or building society savings or investment accounts.

A potential 'danger' is that the valuer can be implying an effective gross return on the sinking fund that is greater than the remunerative rate used in a valuation. Conventionally, a 6% freehold yield might become a 7% leasehold yield, but 7% and 4% with a tax rate of 50p in the £1 would imply that the investor could invest in a sinking fund and initially be better off than by purchasing the leasehold interest because 4% net is 8% gross. It can be seen that the profit rent must grow at a rate sufficient to compensate for the opportunity cost of investing at 7% gross and not at 4% net.

This would be perfectly tenable under certain market conditions.

Both approaches rely on the argument that the purchaser of a leasehold investment is unlikely to have the opportunity to reinvest at the remunerative rate. Apart from the fact that there is no need to reinvest if the investor accepts a return on capital at risk, there is the obvious point that relatively safe rates, albeit variable, are readily available for regular saving - such as government stock at 3.5% to 4.5%.

Further, purchasers of leasehold investments are not purchasing a single, one-off property investment, but are investing in more than one property and more than one type of investment and may receive sufficient income each year to purchase further leasehold investments. The leasehold investor is then better advised to reinvest in further leaseholds (or freeholds) rather than in a sinking fund.

Historically, there may have been some justifications for the dual rate approach. In the eighteenth, nineteenth and early twentieth century leaseholds had long terms (frequently ground leases or improved ground leases with no rent reviews) and were therefore comparable to freehold ground rent investments. Initially they were valued at a single rate at 0.5 to 2% above the freehold rate to reflect the additional risk arising from lease covenants, etc.

The problem emerged in the 1920s and 1930s when lease length began to fall, and when the gap between safe and risky investments became greater. This was in part a reflection of depression in the economy.

It was also a time when purchasers of leases were individuals buying for occupation rather than to hold as investments; there were few large investors in the market. By the mid-1930s, valuers were split between single rate and dual rate, although writers still made the point that sale prices could be analysed on either basis. The need to demonstrate parity with freeholds began to dominate, as did the concern for capital recovery when safe investments were yielding so little.

For those who were seeking protection and perpetuation through reinvestment there was the added difficulty of finding reinvestment opportunities for small annual sums other than in a bank or post office. The result was a growing support for using a dual rate approach.

The position post-1960 was very different, as inflation became part of life. Nevertheless, the dual rate method survived virtually unchallenged other than for the split between the advocates of non-adjusted dual rate and tax adjusted dual rate. It survived in the 1960s, 1970s and 1980s with the low sinking fund even though most savers could achieve much higher returns than the 4% maximum used by valuers. It survived even though few, if any, sinking fund policies were taken out, even though occupiers were depreciating leaseholds as wasting assets in their accounts and even though investors accepted that they effectively provided for capital replacement by reinvestment in a general sense rather than on a per property basis.

A final point is that where the investment is made with borrowed funds or partially borrowed funds (e.g. 60% mortgage), reinvestment at the rates adopted in valuations by certain valuers will certainly be at levels lower than the interest charged on the borrowed capital. Clearly, the investor would be better advised to pay off part of the loan rather than reinvest in a sinking fund.

One of the most disturbing elements of dual rate valuation practice is its concentration on today's profit rent and its failure to consider the future; in particular its failure to reflect the unique gearing characteristics of longer leaseholds. Taking, for example, Property A held for 20 years at £10,000 pa without review and sub-let with five-year reviews for £20,000 pa, and Property B similarly held at £80,000 pa and sub-let at £90,000 pa, the conventional dual rate approach would place the same value on both as they both produce a profit rent of £10,000 pa (see Chapter 6).

This completely hides the fact that with rental growth at any rate per cent, the profit rents after each rent review grow at different rates. This problem of gearing can only be solved by valuing the head rent and occupation rent separately and by adopting a DCF or real value approach. Conventional dual rate methodology would, if applied constantly to every situation, provide the investor with some phenomenal investment opportunities and result in some sellers grossly underselling their investments.

The dual rate method was considered in depth in a RICS research report by Trott (1980) and by Baum and Crosby (2005 and 2008). A general conclusion at that time was that it would continue to be used because 'that is how the market does it'. The position now is that for many it continues to be used because 'that is what they (the valuers) were taught', because 'it continues to be taught because it has an intellectual challenge', and because it seems simple to add 2% to the freehold rate rather than search for comparable leasehold rates.

The authors would maintain that it is defunct for the following reasons:

1. Investors do not seek a return on initial outlay throughout the lease term, they all adopt a portfolio approach and appreciate and use the investment concepts of net present value (NPV) and IRR.
2. Investors have never and still do not reinvest in sinking funds.
3. Investors invest regularly in time-limited investments such as short and medium dated stocks, and do not seek to equate finite investments with infinite (freehold) investments, and do not use dual rate in those markets.
4. Valuation of leaseholds is based on the analysis of freehold sales, or on the basis of analysis of leaseholds based on false assumptions as to tax rates and sinking fund rates.
5. Valuation of anything other than a fixed profit rent produces added issues for the valuer to overcome if the dual rate approach is to be used.

The following section reviews some of the dual rate issues. The authors offer their sympathies to those students who are still required to learn dual rate in all its manifestations.

### Question 1: Dual rate

Value a leasehold profit rent of £20,000 for 4 years using 8% and 3% adj. tax at 40% and prove that a return on capital at 8% and a return of capital at a net rate of 3% is achieved.

## Variable profit rents

Variable profit rents present special problems in valuations where the sinking fund rate of return adopted is lower than the investment risk rate (remunerative rate).

The valuation of a leasehold investment may involve a variable profit rent in the following cases:

1. Where a leaseholder lets property to a sub-lessee for part only of the unexpired term. In this case there will be a fixed profit rent for 'n' years and a reversion to a different level of profit rent in 'n' years.
2. Where a leaseholder lets property to a sub-lessee for the full unexpired term with a proviso for rent reviews in 'n' years to a known sum or to full rental value.
3. Where a leaseholder has a right to a new lease under the Landlord and Tenant Act 1954 Part II, and where the rent for the new lease must be at market rental value adjusted for goodwill and/or approved improvements carried out by the lessee (see Landlord and Tenant Act 1954 Part II and Law of Property Act 1969).

Wherever a leasehold valuation is required the valuer should exercise care in determining the true net profit rent, special attention being given to the allowance for expenditure of repairs and insurance when a sub-letting has taken place on terms differing from the head lease.

In some cases the profit rent will be rising, while in others it will be falling. In both cases anomalies occur if, in the assessment of present worth, the flow of income is treated as an immediate annuity plus one or more deferred annuities, assessed on a dual rate basis. The valuation should be treated as an immediate variable annuity.

Whilst the valuer's subjective adjustments to the remunerative rates used in the three cases may well differ, the particular problem of variable profit rents will remain and is fully illustrated in the following problem.

### Example A.5

Using as an example a profit rent of £1,800 for two years rising to £2,000 for five years, demonstrate the problems encountered when valuing rising profit rents on a conventional dual rate basis. Demonstrate the alternatives to this approach.

Leasehold rate at market rent: 10% and 3% adj. tax @ 40%

Conventional valuation:*

*Term:*

| | | |
|---|---|---|
| Profit rent | £1,800 | |
| YP for 2 years at 9% and 3% adj. tax @ 40% | 1.0977 | £1,976 |

*Reversion:*

| | | |
|---|---|---|
| Profit rent | £2,000 | |
| YP for 5 years at 10% & 3% adj. tax at 40% | 2.416 | |
| | £4,832 | |
| PV £1 in 2 years at 10% | 0.8265 | £3,994 |
| | | £5,970 |

*The use of variable remunerative rates causes further problems, as does the deferment of the reversions at the reversionary remunerative rate. These issues can be overcome by adopting a same yield approach.

### What are the problems?

None is apparent from this valuation, but compare the valuation of the following profit rent of £1,800 receivable for seven years (five years plus two years):

| | |
|---|---|
| Profit rent | £1,800 |
| YP for 5 years at 10% and 3% adj.tax at 40% | 3.1495 |
| | £5,669.10 |

Notice that the second investment is valued at £300 lower than the first.

But what is the real difference?

The first investment produces an extra £200 for five years deferred for two years.

| | |
|---|---|
| Profit rent | £200 |
| YP for 5 years at 10% and 3% adj. tax at 40% | 2.416 |
| | £483.2 |
| PV £1 in 2 years at 10% | 0.8265 |
| | £400.0 |

The difference in the two valuations is £300, yet the difference in rent is worth £400. There is obviously some kind of error. The error can be demonstrated in the same manner much more dramatically.

Take two profit rents with a 20-year life.

One is of £1,000 for the whole period; the other rises to £1,100 after 10 years (and is therefore more valuable).

| (1) | | (2) | | |
|---|---|---|---|---|
| Profit rent | £1,000 | Profit rent | £1,000 | |
| YP for 20 years at 10% and 3% adj. tax at 40% | 6.1718 | YP for 10 years at 9% and 3% adj. tax at 40% | 4.2484 | £4,248 |
| | £6,172 | Reversion to | £1,100 | |
| | | YP for 10 years at 10% and 3% adj. tax at 40% × PV £1 in 10 years at 10% | 1.573 | £1,730 |
| | | (4.0752 × 0.386) | | £5,977 |

The inferior investment is valued more highly by the conventional dual rate method, and the error could be embarrassing because it generally remains unnoticed by most valuation surveyors.

It can be demonstrated, without any need for comparison, that an error does exist, by checking that the sinking fund actually replaces the initial outlay. This will be done with reference to the original valuation.

Term

| | |
|---|---|
| Capital value | £5,970 |
| Income | £1,800 |

Spendable income

0.09 × £5,970

= £537.3

Sinking fund

£1,800 – £537.3

= £1,262.7 gross

= 0.6 (£1,262.7) net

= £757.6

A £1 pa for 2 years at 3%

2.03

£1,538

This £1,538 will then be allowed to accumulate interest for five more years, the period of the reversion.

£1,538

A £1 in 5 years at 3%

1.1593 = £1,783 replaced

*Reversion*:

| | |
|---|---|
| Capital value | £5,970 |
| Income | £2,000 |

Spendable income

0.10 × £5,970

= £597

Sinking fund

£2,000 – £597

= £1,403 gross

= 0.6 (£1,403) net

= £842

A £1 pa for 5 years at 3%

5.3091

£4,470 replaced

Total replacement of capital

£1,783 + £4,470 = £6,253

Compare £6,253 with an initial outlay of £5,970 and it can be seen that there is an over-replacement of capital.

### Why?

It can be shown that the replacement of capital for both term and reversion, when examined separately, is perfectly correct.

*Term:*

| | |
|---|---|
| Capital value | £1,976 |
| Income | £1,800 |

Spendable income

0.09 x £1,976

Sinking fund

£1,800 – £178 =

£1,622 gross

= £178

× 0.6

£973.2 net

A £1 pa for 2 years at 3%

2.03

Term capital replaced £1,976

*Reversion:*

| | |
|---|---|
| Capital value | £4,832 |
| Income | £2,000 |

| Spendable income | Sinking fund | |
|---|---|---|
| 0.10 × £4,832 | £2,000 – £483.2 | = £1,516.8 gross |
| = £483.2 | | × 0.6 |
| | | £910.08 net |
| | A £1 pa for 5 | |
| | years at 3% | 5.3091 |
| | Reversion capital replaced | £4,832 |

It follows that the error must arise from the addition of term and reversion. This produces an extra accumulation of sinking fund resulting in an over-replacement of capital. The deferment of the reversion by a single rate PV is another expression of the same error. It is often said that the error results from the provision for two sinking funds or the interruption of the desired single sinking fund. As a result the methods that have been devised to deal with this error attack the problem by attempting to ensure that the initial capital is accurately replaced by the sinking fund.

## Method 1: The sinking fund method

The problem which has been identified above is that the conventional dual rate method of valuing leasehold interests does not provide for accurate replacement of capital. The sinking fund method ensures that this must happen, its premise being that capital value is equal to the amount replaced by the sinking fund. The method calculates the amount of the net sinking fund and its accumulation which must necessarily be equal to the capital value of the investment.

Let capital value = $x$

Term:

| | |
|---|---|
| Rent | = £1,800 |
| Gross sinking fund | = Income - spendable income |
| | = £1,800 - 0.09$x$ |

(Return on capital = remunerative rate of 9%: return on capital = 0.09$x$)

Net sinking fund = (£1,800 – 0.09$x$) (0.6) = £1,080 – 0.054$x$

Reversion:

| | |
|---|---|
| Rent | = £2,000 |
| Gross sinking fund | = £2,000 – 0.10$x$ |

Therefore:

| | |
|---|---|
| Net sinking fund | = £1,200 – 0.06$x$ |

Calculate accumulation of net sinking funds:

| Term: | | £1,080 – 0.054x | |
|---|---|---|---|
| A £1 pa for 2 years at 3% | 2.03 | | |
| A £1 in 5 years at 3% | x 1.1593 | 2.3534 | £2,541.65 – 0.127x |

| Reversion: | | £1,200 – 0.06x | |
|---|---|---|---|
| A £1 pa for 5 years at 3% | | 5.3091 | £6,370.29 – 0.31285x |

Adding the term and reversion equations together £8,912.57 – 0.4455x

The capital value should equal the amount replaced and therefore $x$ should equal the sum of the sinking fund accumulations.

So:

$$x = £8,912.57 - 0.4455x$$
$$1.4455x = £8,912.57$$

Therefore:

$$x = £6,165.74$$

This method can be checked by checking the accumulation of sinking funds on term and reversion.

Term:

| Capital value | £6,166 | | |
|---|---|---|---|
| Income | £1,800 | | |

| Spendable income | Sinking fund | | |
|---|---|---|---|
| 0.09 x £6,166 | £1,800 – £554,94 | = £1,245.06 gross | |
| = £554.94 | | × 0.6 | |
| | | £747.04 net | |
| A £1 pa for 2 years at 3% | 2.03 | | |
| A £1 in 5 years at 3% | 1.1593 | 2.354 | £1,758.07 |

Reversion:

| Capital value | £6,166 | | |
|---|---|---|---|
| Income | £2,000 | | |

| Spendable income | Sinking fund | | |
|---|---|---|---|
| 0.10 × £6,166 | £2,000 – £616.6 | = £1,383.40 gross | |
| = £616.60 | | × 0.6 | |
| | | £830.06 net | A |
| £1 pa for 5 years at 3% | | 5.3091 | £4,406.87 |
| Total capital replaced (£1,758.07 + £4,406.87) = | | | £6,164.94* |

*The marginal error here is due to the initial rounding to £6,166 and subsequent rounding in the calculations.

## Method 2: The annual equivalent method

The purpose of this second method is to find that fixed income which would be equivalent to the rising profit rent which is to be valued. The current teaching of this and the sinking fund method is attributed to Dr MJ Greaves previously of Reading University and the National University of Singapore.

Equivalent incomes for both term and reversion are found and valued separately to allow for the use of different remunerative rates on term and reversion.

It was originally suggested that the rate of interest used to capitalise and de-capitalise both incomes when finding the annual equivalent should be the accumulative rate, this approach is adopted in the following example:

---

**Example A.6**

*A Capitalisation at low safe rate*

Term:

| | | |
|---|---|---|
| Income | £1,800 | |
| YP for 2 years at 3% | 1.9135 | £3,444 |

Reversion:

| | | | |
|---|---|---|---|
| Income | | £2,000 | |
| YP for 5 years at 3% | 4.5797 | | |
| PV £1 in 2 years at 3% | × 0.9426 | 4.3168 | £8,633 |

*B Find annual equivalent income*

Term:

$£3,444 \div$ YP for 7 years at 3%

$= £3,444 \div 6.2303$

$= £552.78$

Reversion:

$£8,633 \div$ YP for 7 years at 3%

$= £8,633 \div 6.2303$

$= £1,385.65$

*C Capitalise annual equivalents at market capitalisation rate*

Term:

| | | |
|---|---|---|
| Income | £552.78 | |
| YP for 7 years at 9% and 3% adj. tax at 40% | 3.2519 | £1,797.59 |

Reversion:

| | | |
|---|---|---|
| Income | £1,385.65 | |
| YP for 7 years at 10% and 3% adj. tax at 40% | 3.1495 | £4,364.10 |
| | | £6,161.69 |

### D Proof

Term:

| | |
|---|---|
| CV | £1,797.59 |
| Income | £1,800 |
| Spendable Income | Sinking fund |

$0.09 \times £1,797.59$     $£1,800 - £598.19^* =$   £1,201.81 gross

$= £161.78$                                  $\times 0.6$

                                               £721.09 net

*Reversion*:

| | |
|---|---|
| Capital value | £4,364.10 |
| Income | £2,000 |
| Spendable income | Sinking fund |

$0.10 \times £4,364.10$     $£2,000 - £598.19^* =$   £1,401.81 gross

$= £436.41$                                  $\times 0.6$

                                         £841.09 net

*Total spendable income at 9% and 10% on term and reversion is £598.19 (£161.78 + £436.41).

Check sinking fund accumulations:

| Term: | | £721.09 |
|---|---|---|
| A £1 pa for 2 years at 3% | 2.03 | |
| A £1 in 5 years at 3% | $\times 1.1593$ | 2.354 |
| | | £1,697 |

| Reversion: | | £841.09 |
|---|---|---|
| A £1 pa for 5 years at 3% | | 5.3091    £4,465 |
| | | £6,162* |

*Error due to rounding.

The proof employed in the sinking fund approach is not applicable in this case, where term and reversion must be kept separate when checking the sinking fund accumulation. In the first approach, spendable income differs over term and reversion; but in the annual equivalent method, spendable income remains the same. If the remunerative rates were the same over term and reversion then either proof may be used, but when they differ; the annual equivalent valuation can only be checked by the particular approach outlined above.

## Method 3: The double sinking fund method

This, the original of the three methods discussed here, involves more detailed arithmetic. The required sinking fund to replace a capital value of x is deducted from income to leave the amount of spendable income. The spendable income is then capitalised. This ignores the sinking fund accumulation, which is then added back to produce a similar equation to that which is solved in the sinking fund approach.

A constant sinking fund is ensured by this approach, overcoming the conventional method's fault. This method is attributed to AW Davidson, one time Head of Valuation at the University of Reading.

1. Let capital value = x Then:

| | |
|---|---|
| sinking fund to replace x | = ASF to replace £1 in 7 years at 3% × x |
| | = 0.1305064x |

Gross up for tax at 40%

$$= 0.1305064 \times \left( \frac{1}{1-t} \right)$$

$$= 0.1305064x \times (1/(1-0.06))$$

$$= 0.2175106x$$

Therefore:

Spendable income is £1,800 − 0.2175106x for term.
Spendable income is £2,000 − 0.2175106x for reversion.

2 Capitalise the spendable income

Term:

| Income | £1,800 − 0.2175106x | |
|---|---|---|
| YP for 2 years at 9% | 1.7591 | £3,166.38 − 0.3826228x |
| Reversion: | | |

| Income | £2,000 − 0.2175106x | |
|---|---|---|
| YP for 5 years at 10% | 3.7908 | |
| PV £1 in 2 years at 10% | 0.8264 | 3.133 |

£6,265.79 − 0.6814372x

Adding the term and reversion

£9,432.17 − 1.06406x

This (£9,432.17 minus 1.06406x) is the PV of the spendable income provided by the investment. It has been capitalised by a single rate YP which contains an inherent sinking fund at the remunerative rate. The capital value of the spendable income could thus be reinvested at the end of seven years, while another sinking fund has been provided to replace the capital value of the whole income flow - x. There are therefore two sinking funds, hence the name of the method.

An alternative view is to remember that income is split into two parts when it is received for a limited period - spendable income and sinking fund. It follows that capital value can be split into capitalised spendable income and capitalised sinking fund. Having found the first of these constituents the second should be added to give the total value of the investment.

What, then, is the present capital value of the sinking fund?

It replaces x in seven years.

Its present value is x deferred for seven years at the investor's remunerative rate(s): 9% for two years and 10% for the remaining five.

3. Replaced capital value =x

| PV £1 in 5 years at 10% | 0.6209213 | |
|---|---|---|
| PV £1 in 2 years at 9% | 0.84168 | 0.522617 |

0.522617x

Thus if x is the capital value CV, then x must be equal to the total of 1, 2 and 3.

$$x = £9,432.17 - 1.06406x + 0.522617x$$
$$x = £9,432.17 - 0.541443x$$
$$1.541443x = £9,432.17$$
$$x = £6,119.05$$

Proof:

Term:

| | |
|---|---|
| CV | £6,119.05 |
| Income | £1,800 |

| Spendable income | Sinking fund | |
|---|---|---|
| $0.09 \times £6,119.05$ | £1,800 – £550.71 = | £1,249.29 gross |
| = £550.71 | | $\times 0.6$ |
| | | £749.57 net |

*Reversion:*

| | |
|---|---|
| CV | £6,119.05 |
| Income | £2,000 |

| Spendable income | Sinking fund | |
|---|---|---|
| $0.10 \times £6,119.05$ | £2,000 - £611.91 = | £1,388.09 gross |
| = £611.91 | | $\times 0.6$ |
| | | £832.85 net |

Check sinking fund accumulations

| | | | |
|---|---|---|---|
| Term: | | £749.57 | |
| A £1 pa for 2 years at 3% | 2.03 | | |
| A £1 in 5 years at 3% | x 1.1593 | 2.354 | £1,764.02 |
| Reversion: | | £832.85 | |
| A £1 pa for 5 years @ 3% | | 5.3091 | £4,421.68 |
| | | | £6,185.70 (compare with CV of £6,119.05) |

The proof suggests that capital value is not accurately replaced. However, the rationale of the method must ensure accurate replacement of capital, and this leads to the conclusion that this method suffers from another fault. This is that although rates of return of 9% on the term and 10% on the reversion are required, they are not accurately provided by this valuation.

There is an over-replacement of capital and consequently the interest is undervalued, as in the conventional method, but to a reduced degree.

| | CV | Replaced CV |
|---|---|---|
| Conventional dual rate | £5,970 | £6,253 |
| Sinking Fund | £6,166 | £6,165 |
| Annual equivalent | £6,162 | £6,162 |
| Double sinking fund | £6,119 | £6,186 |

It is therefore possible to conclude that the sinking fund and annual equivalent methods best overcome the problem posed by the conventional method of valuing variable profit rents as they appear to provide for accurate replacement of capital and correctly allow for the required rate of return to be provided.

However, even the apparent accuracy of these solutions cannot be relied upon in all circumstances. For example, the sinking fund method becomes unworkable if the term income is particularly low; the spendable income over the period becomes negative, and considerable problems arise.

Harker, Nanthakumaran and Rogers (1988) of Aberdeen University support the sinking fund method and suggest a resolution to the negative spendable income problem.

A popular alternative to these methods is the Pannell method. In this method the capital value of the variable profit rents are found by capitalising each on a single rate basis using the appropriate leasehold remunerative rates. The annual equivalent of the product is then found by dividing through by the YP for the full unexpired term at the remunerative rate or the average of the rates used. This annual equivalent can then be capitalised on a dual rate basis for the full term.

As the use of dual rate valuations can be seen to lead to difficulties, single rate leasehold valuations are recommended. The fact that few valuations of variable profit rents are actually valued using one of the corrective approaches raises other questions about the validity of using dual rate.

It should be noted that if single rate is used the valuations may be prone to error where more than one rate of interest is used. Nonetheless, the potential for arithmetical error in a single rate valuation is considerably less. A reasonable conclusion to this discussion on dual rate is that if valuers are supposed to mirror the market in the selection of valuation methods then it is reasonable to suggest that if buyers in the market do not use dual rate then neither should valuers. Most of this is therefore redundant, but hopefully provides for interesting reading.

---

**Question 2: Dual rate and DCF**

A shopping centre produces rents of £2.5 million with rents due for review in 3 years to £3.25 million and thereafter every 5 years. The centre is held leasehold with 60 years of the lease to run with one rent review of the ground rent in 10 years time. The ground rent is currently £50,000 and the rent review is fixed at 5% of the shop rents.

Set out a skeleton valuation of the head lease using a dual rate approach at 8% and 3% with tax at 40p and a DCF approach on the basis of a freehold target rate of 10%.

## Tax adjustment rate

What rate of tax should be used to adjust a gross profit rent to a net rent, or to gross-up a sinking fund to allow for tax?

The answer normally given is to use standard rates of income or corporation tax but the problem is more complex due to the presence of non-taxpayers such as pension funds and charities. The valuer's task is generally to determine market value, this implies a sale to the most probable purchaser, which implies some knowledge of the market and therefore sufficient knowledge to adjust, both for analysis and valuation, at a rate appropriate to the most probable purchaser.

However, as tax is levied at different rates for different investors, an average tax rate (e.g. 40%) is used to reflect market interaction. However, analysis of leasehold investments from a client's point of view must be carried out at that client's net of tax rate.

### Example A.7

'Short leasehold investments are sound investments for charities because of their tax advantage.' Discuss.

Apart from the point that charities are more interested in income over the long term than the short term, a number of points can be made. If the statement is true then charities must comprise a sub-market for this kind of investment. In that case the valuer needs to reflect the fact that the income is probably tax free; that they would be ill-advised to reinvest in a sinking fund policy with an assurance company because they will probably have difficulty in recovering the tax element on the 4% net accumulations of the policy; and that they would be best advised to reinvest in the safest gross funds to avoid delay in recovery of taxed interest or dividends.

### Example A.8

Analyse the sale of a £100 profit rent for four years at a price of £200 from the point of view of a gross (ie non taxpaying) fund.

$$£200/£100 = \text{multiplier of } 2$$

(a) Assume reinvestment in an equivalent safe gross fund at, say, 8%. Therefore the sinking fund to recover £200 will be:

$$£200 \times \text{ASF to replace £1 in 4 years at 8%}$$
$$= £200 \times 0.22$$
$$= £44$$

The gross spendable income = £100 − £44
$$= £56$$

Therefore

The gross return        $= (£56/£200) \times 100 = 28\%$

(b) Without a reinvestment assumption analyse to find the IRR.

$$\text{Solve}: \quad \frac{1 - \left(\dfrac{1}{(1+i)^4}\right)}{i} = 2$$

The present value of £1 pa in four years at 35% = 2, and therefore the IRR is 35%.

Charities, because of their tax position, may be interested in short leasehold investments, because, if they can buy at prices determined by conventional dual rate approaches, the effective return will be sufficiently high to compensate for the disadvantages of the investment.

## Summary

The historical evolution of the dual rate method is now shrouded in the mists of time. Mackmin (2008) provides a reasoned explanation for its development based on a review of historic articles, books and letters. Textbooks and journals of the time move from single rate to dual rate with very little explanation and with virtually no consideration of the fact that it hinges on the acceptance of a return on initial outlay throughout the life of the investment, a concept apparently unique to the UK leasehold property investment market.

The emergence of the tax adjustment factor is better documented and is relevant to all short-term investments, however valued, where tax is charged on that element of income which is essentially capital recoupment.

Part of the market is still as reluctant today to return to single rate as it was initially to shift from single rate to dual rate. One of the supposed strengths of dual rate lay in the ability to compare returns from freeholds with leaseholds. However, this is fallacy as the comparison is one between an Internal Rate of Return (IRR) and a Sinking Fund Return (SFR). The reluctance today may be the fear of the unknown arising from the unique nature of every leasehold investment and the pressure on the valuer to find a unique solution. The solution lies in the proper use of discounting techniques where the unique growth expectations can be explicitly accounted for.

In summary, dual rate methodology can be criticised for the following reasons:

- Sinking funds per se are not available in the investment market.
- Leaseholds are not directly comparable to freeholds and remunerative rates cannot be taken as simply 1% to 2% above freehold rates.
- Accumulative rates do not appear to be market sensitive. 4% has been used when safe rates have been as high as 12% and when they have been below 4%.
- It is not possible to derive a remunerative rate for dual rate or dual rate adjusted for tax from market sales unless the accumulative rate and the tax rate are both assumed by the valuer.
- Dual rate analysis of variable geared profit rents is similarly impossible.
- There is an arithmetic error in dual rate when used for valuing variable profit rents.

# Spreadsheet user

## Project 1 : Dual rate valuation

Using Excel, create an area for entry of the variables and display of the answers. An example is indicated below:

|   | A | B | C | D | E | F |
|---|---|---|---|---|---|---|
| 1 | Valuation Spreadsheet to calculate the capital value of a terminable income which in taxable | | | | | |
| 2 | | | | | | |
| 3 | | | | | | |
| 4 | ENTER THE FOLLOWING VALUATION DATA: | | | | | |
| 5 | | | | | | |
| 6 | | | | Income £ | | |
| 7 | | | | Period ($n$) (years) | | |
| 8 | | | | SF Rate ($i_s$) % | | |
| 9 | | | | Interest Rate ($i$) % | | |
| 10 | | | | Tax Rate ($t$) % | | |
| 11 | | | | | | |
| 12 | | | | CAPITAL VALUE | | |

In cell F12 you will need the formula for dual rate YP adjusted for tax which can be expressed with reference to the above cell layout as:

$$E6*(1/(E9 + (((E8/(((1+E8)^{\wedge}E7)-1))*((1/(1-E10))))))))$$

Note: The above assumes that the cells E8, E9 and E10 are in a percentage format (see spreadsheet user, Project 1); if not the values in the above formula require division by 100.

Your capital value calculator spreadsheet should look like the following screenshot:

Microsoft Excel - Book1

File  Edit  View  Insert  Format  Tools  Data  Window  Help

Arial          ▾ 10 ▾  B  I  U  ≡ ≡ ≡ ▦  🔲 % ,  ⁺⁰ ⁰⁰  ⏐ ⏐

G16          ▾       fx

|   | A | B | C | D | E |
|---|---|---|---|---|---|
| 1 | Valuation Spreadsheet to calculate the capital value of a terminable | | | | |
| 2 | income which is taxable | | | | |
| 3 | | | | | |
| 4 | ENTER THE FOLOWING VALUATION DATA : | | | | |
| 5 | | | | | |
| 6 | | | | Income            £ | 2000 |
| 7 | | | | Period (*n*) (years) | 10 |
| 8 | | | | SF Rate (*i*s)      % | 3.000% |
| 9 | | | | Interest Rate (*i*)  % | 10.000% |
| 10 | | | | Tax Rate  (*t*)     % | 25.000% |
| 11 | | | | | |
| 12 | | | | CAPITAL VALUE | 9246.1032 |
| 13 | | | | | |

# Appendix B

## Illustrative Investment Property Purchase Report

The report set out here is based on material originally provided by Gooch and Wagstaff, Chartered Surveyors, for the third edition of this book. It has been regularly amended by the authors, as too have the reporting requirements of the Royal Institution of Chartered Surveyors (RICS) and the expectations of clients. It is used here solely as an illustration for students who are sometimes asked on their courses to write valuation reports and have very little material to access. We hope it helps but do not grade it in academic or practice terms. We also appreciate sight given to us by DTZ Debenham Tie Leung Ltd of a more recent RICS Valuation Standards (The Red Book) report.

A full investment report, complete with all background information and descriptive material, may well run to 50 pages and longer for the more complex multi-million pound, multi-tenanted properties.

The best initial guide as to content is the requirement set out in The Red Book Valuation Standards (VS) 6.1 (italics are RICS' italics and indicate words defined in The Red Book glossary or elsewhere):

The report must clearly set out the conclusions of the valuation in a manner that is not ambiguous, misleading, or create a false impression. It must also deal with all the matters agreed between the client and the valuer in the *terms of engagement* and include the following minimum information, except where the report is to be provided on a form supplied by the client:

(a) identification of the client;
(b) the purpose of the valuation;
(c) the subject of the valuation;
(d) the interest to be valued;
(e) the type of property and how it is used, or classified, by the client;
(f) the *basis*, or *bases*, of *value*;
(g) the date of valuation;
(h) disclosure of any material involvement or a statement that there has not been any previous material involvement;
(i) if required, a statement of the status of the valuer;
(j) where appropriate, the currency that has been adopted;
(k) any *assumptions, special assumptions*, reservations, any special instructions or *departures*;

(l)  the extent of the valuer's investigations;
(m) the nature and source of information relied on by the valuer;
(n)  any consent to, or restrictions on, publication;
(o)  any limits or exclusion of liability to parties other than the client;
(p)  confirmation that the valuation accords with these standards;
(q)  a statement of the valuation approach;
(r)  a statement that the valuer has the knowledge, skills and understanding to undertake the valuation competently;
(s)  the opinions of value in figures and words;
(t)  signature and *date of the report*.

**Source: RICS Valuation Standards 6.1.**

The report must also contain the previously agreed terms of engagement and must include appropriate commentary on any specific matters referred to in those terms. VS2.1 specifies these terms:

Confirmation of terms of engagement.

The *terms of engagement* provided in compliance with VS 1.4 must be in writing and, at a minimum, include the following terms:

(a)  identification of the client;
(b)  the purpose of the valuation;
(c)  the subject of the valuation;
(d)  the interest to be valued;
(e)  the type of property and how it is used, or classified, by the client;
(f)  the *basis, or bases, of value*;
(g)  the date of valuation;
(h)  disclosure of any material involvement, or a statement that there has not been any previous material involvement;
(i)  if required, a statement of the status of the valuer;
(j)  where appropriate, the currency to be adopted;
(k)  any *assumptions, special assumptions*, reservations, any special instructions or *departures*;
(l)  the extent of the valuer's investigations;
(m) the nature and source of information to be relied on by the valuer;
(n)  any consent to, or restrictions on, publication;
(o)  any limits or exclusion of liability to parties other than the client;
(p)  confirmation that the valuation will be undertaken in accordance with these standards;
(q)  confirmation that the valuer has the knowledge, skills and understanding to undertake the valuation competently;
(r)  the basis on which the fee will be calculated;
(s)  where the *firm* is registered for regulation by RICS, reference to the *firm's* complaint handling procedure, with a copy available on request;

(t) a statement that compliance with the standards may be subject to monitoring under the institution's conduct and disciplinary regulations.

**Source: RICS Valuation Standards VS 2.1**

The terms of engagement may well be contained in an appendix to the main report. Appendix 6.1 of The Red Book provides a student-friendly table with comments relating to each aspect of the report.

# The report

The report will normally begin with an executive summary of the salient points, but with a proviso (see below) and may take the following form. Such reports will be fully supported with photographs, location plans, floor plans with measurements as per the RICS Code of Measuring Practice, market evidence; in other words it will take the reader to the logical conclusion arrived at by the valuer and will not be ambiguous, misleading, or create a false impression.

1.0 Executive summary

Terms have been agreed on behalf of XYZ pension fund to purchase the freehold shop investment known as 20 Market Street, Market Town, AB1 2CD from (name and address of the current owner) for £525,000 (five hundred and twenty five thousand pounds). In brief:

| | |
|---|---|
| Tenure: | Freehold |
| Net Internal Area: | 100 m$^2$ |
| User: | High quality retail outlet |
| Tenancy: | Let to AAA and BBB trading as ZZZ bakers on an assignment from XXXX(UK) Ltd for a term of 25 years from 00/00/00 to 00/00/00 on FRI terms at £24,000 a year payable quarterly in advance on the normal quarter days |
| Rent reviews: | every five years upward only the next review being on 00/00/00 |
| Estimated market rent: | £30,000 a year on similar terms |
| Estimated yields: | 4.32% net initial rising to 5.40% on reversion in 00/00/00; net equivalent yield 5.27%. |

Our conclusion is that the price agreed reflects the current market and we recommend the acquisition at this price to XYZ pension fund.

This summary must only be read in conjunction with the detailed comments which follow.

2.0 Client

This report has been prepared in accordance with the agreed terms and conditions as set out in annex 1 for the XYZ pension fund, Fund House, Edinburgh, Scotland, E1 1AZ.

3.0 Valuation purpose

The purpose of this valuation is to advise XYZ on the acquisition of the freehold interest in 20 Market Street, Market Town, AB1 2CD at an agreed price of £525,000 (five hundred and twenty five thousand pounds) subject to contract and survey. It should be noted that whilst we have inspected the property for the purpose of the valuation we are not instructed to report on its structural condition. We understand that a full survey has been carried out on your behalf by........................................... and that you are satisfied with the result. Further that you have commissioned an environmental report which we have seen and we note that it has paid full regard to the RICS guid-ance note *Contamination, the environment and sustainability; their implications for chartered surveyors* 3e. Any aspects of this report relevant to our valuation our noted in section 17.0.

4.0 Subject matter

This report relates solely to the property known as 20 Market Street, Market Town AB1 2CD as identified on location plan in annex 2.

5.0 Interest

The interest to be valued is the freehold, Land registration title No.12345 being offered for sale subject to the lease dated 00/00/00 as detailed in section 14. We understand from your solicitors, S S and S, that there is a right of way in fee simple to the owners and occupiers of the adjoining property, 22 Mar-ket Street, over the passageway along the side of the subject property. This gives access to the upper floors of the adjoining property. This is marked in green on the plan in annex 2.

## 6.0 Description and construction

The property is a Grade II listed building constructed circa 1850 comprising a ground floor shop unit with single storey rear storage and three floors of disused residential space above having access from a rear yard. Access to the yard is provided via a passageway to the side of the shop (see 5.0 above). The shop has limited internal width averaging 3.2 metres.

The building is constructed of brick and sandstone quoin blocks and eaves cornice under a pitched slate roof. The windows are single glazed wooden sash. The shop has no central heating but is warmed by wall-mounted electric heaters. The upper floors are heated by a coke burning boiler serving radiators, this boiler and the radiators are in need of replacement.

The separate access to the upper floors means that the three floors of disused space could be brought back in to use subject to any necessary consents. Similar space in 22 Market Street was recently converted to office use.

## 7.0 Basis of value

The basis of value is market value.

> The RICS Red Book definition will be included here or in an annex together with the conceptual framework as published in International Valuation Standard 1 and reprinted in the Red Book.

## 8.0 Date of Valuation

The property has been valued as at 00/00/00. Values may change over time particularly at the present time with difficult borrowing conditions.

## 9.0 Material involvement

Neither the valuer, whose signature appears below nor BBBB, Chartered Surveyors have had any material involvement with this property or its owners.

## 10.0 Status of the valuer

> In certain circumstances valuers will need to declare whether they are acting as an internal or external valuer; this might be the case here if this report was being prepared by a valuer who is an employee of the pension fund.

## 11.0 Currency

> It is sometimes necessary to state the currency of the valuation and any exchange rates that might have been used to express the value in another currency. This would not be needed for a UK property with a UK valuer and client.

## 12.0 Assumptions and special assumptions

> If assumptions have been agreed with the client as part of the terms of engagement they must be restated in the report.

## 13.0 Investigations

## 13.1 Location

The property is located in an historic City with a population of 100,000 persons which is substantially increased by tourists and students. The estimated shopping catchment is some 200,000 persons (see annex c). The City is situated close to the junction of the MO and AOO this providing good access to London (50 miles), XXX (00 miles) and YYY (00 miles). There are good road connections to all parts of the region.

A main line rail service operated by TTTTTTT connects to London King's Cross (the fastest journey time being 1 hr 50 minutes). The Market Town international airport is some 20 minutes by road from the City centre and provides regular flights to Paris, Brussels and Amsterdam but is primarily a holiday destination airport with flights by UUU; PPP, and EEE.

The property is located within the established shopping area in the City centre on the south side of Market Street close to all the major retailers (see Goad plan annex 3).

> An opinion as to the suitability of the location for the current use of the building will be given. In the case of retailing, details of pedestrian flows, travel time catchment, spending power and competitive centres might be included. This is factual information fully sourced.

## 13.2 Site

The site slopes up to the rear and is roughly rectangular in shape. It has a frontage to Market Street of 4.35 metres and a depth of 31.6 metres. The site extends to an area of approximately 126 square metres and is outlined in red on the attached Ordnance Survey extract (annex 2).

## 13.3 Accommodation

> A full breakdown of the accommodation with areas needs to be provided in table form. Metric measures should be used but imperial measures are still in use and frequently it will be imperial followed by metric in brackets. Some thought should be given to the client and there likely expectations e.g. if not UK or US clients then metric first. All measurement should be confirmed as having been in accordance with the RICS Code and floor plans should be provided. The instruments used to measure are often mentioned with a confirmation of any recent calibration.

### 13.4 Services

Details of all services should be given. These will not have been tested and a statement to that effect should be given. Details of the energy certificate will be included and any other related matters.

### 13.5 Town planning

This is a critical section especially for retail, where the permitted use class can have a significant bearing on lettability and market rents. In this case (see 5th edition) the planning position was very complex and was potentially affected by plans for a new retail scheme. Anything relevant to the trading potential of a retail unit must be discovered and noted. This can include highway changes, car park changes, pedestrianisation.

### 13.6 Rateable Value

We have checked the RV on the Valuation Office website and have noted that this property is listed with an RV (2010) of £28,000. The current Uniform Business Rate is 00.0p in the £. The tenant is liable for paying the UBR.

### 14.00 Tenancy

The entire property is held on a lease granted in 00/00/00 to XXXX for a term of 25 years ending on 00/00/00. It is subject to upward only rent reviews every five years; the next review is due in 00/00/00. The lease is on full repairing and insuring terms. We understand that you are satisfied that your survey has confirmed the premises are being properly maintained and decorated. This lease was formally assigned with the existing owners consent in 00/00/00 to the current occupiers. The upper floors are included in the lease but are no longer used for residential purposes. In terms of our valuation we have included this space as ancillary to the retail use and not as unimproved residential.

Here or later the valuer needs to confirm whether this information has been taken from the lease or provided by the client's solicitor. Attention should be drawn to key covenants in the lease, to any licenses to carry out alterations, details of contracting out or not from the Landlord and Tenant Act 1954 Part II.

### 15.00 Sources of information

All information necessary for a valuation of this nature have been obtained and there sources are detailed in annex 4.

The valuer must use official sources for details of planning consents and of any tenancies, restrictive covenants etc.

## 16.00 Market rent

Market Town is a cathedral City with a substantial catchment area and is much sought after by national multiple retailers. We are currently aware that AA, BB, CC and D are, even in the current weak market, seeking shop units in Market Street. Market rents have remained stable over the last three years having dipped from their peak in 2007. Relevant extracts from our retail report for this year are included in annex 5 together with evidence of recent lettings and market rents agreed at rent review and lease renewal. Whilst not conclusive in todays uncertain market we feel that there is sufficient evidence to support a zone A rent of £400 m$^2$ which suggests a current market rent in the region of £30,000 for the premises. This might be increased if planning consent could be obtained for a change of use of the upper floors to offices.

## 17.00 Valuation

Our valuation has been prepared in accordance with the sixth edition of the RICS Valuation Standards. The evidence of current market rents suggests a market rent of £30,000.

The market at the time of drafting this 6$^{th}$ edition is very different to that for the 5$^{th}$ edition and good market sales data is hard to find. We have not amended the value given in the 5$^{th}$ edition but would point out that this section in 2011 would need to contain as much evidence as is available and clear warnings as to the level of uncertainty. An investment sale of this nature would still attract interest and in this value range money is available from various individual funds, small property companies and the like; but the statement on market value and yields provided here, whilst probably not far out in 2006 are likely to be 1.5% to 2.0% higher in 2011 suggesting a value closer to £450,000.

Terms have been agreed for the purchase of the freehold interest in 20 Market Street for £525,000 (five hundred and twenty-five thousand pounds) subject to contract. Our own valuation set out in annex 6 using the ABC investment software supports this sum and in our opinion the market value of the freehold interest, subject to the right of way and tenancy is £525,000 (five hundred and twenty-five thousand pounds).

A purchase at this price will give an initial yield of 4.32% rising to 5.40% on reversion in 00/00/00. This represents a net equivalent yield of 5.27%. These yields are after allowance for acquisition costs of 5.75% to cover fees and SDLT.

18.00 Exclusions of liability and publication
This valuation has been prepared in accordance with the terms of engagement agreed on 00/00/00 and set out in annex 1 together with an agreed statement on publication and liability.

> Where it is known that the client needs to include the valuation in a publication then a statement should be provided as a separate document.

The property was inspected and valued by Mr A Proff of Proff and Partners, Chartered Surveyors. A. Proff is a Fellow of The Royal Institution of Chartered Surveyors with over twenty years experience of retail property in the local and national market and has sufficient knowledge, skill and understanding to undertake this valuation.
19.00 Signature
A. Proff.
00/00/00
Appendices as referred to in the text above.

# Additional information which may be required by some investor clients

When advising investors on specific property investment opportunities it is normal practice, in agreement with the client, to provide analysis which goes beyond the basic information on initial, reversion and equivalent yields. The following section describes the nature of the analysis that might be requested and illustrates some of the related calculations. The case used is based on a 2005 transaction; the details have been amended but not updated.

## Background

An institutional client holding a large portfolio is seeking to achieve growth in its revenues, which are almost wholly in the form of rents, in excess of inflation. It must certainly achieve returns in excess of the redemption yield on government bonds, which are risk free and liquid.

It has significant cash funds to invest, is seeking to invest in non-traditional (offices and retail) sectors and the market is very competitive.

## The investment

It has been offered a 41 acre specialist storage site for £29 million. It is occupied by a subsidiary of a well known global brand on a remaining 15-year term with rents rising at a fixed 3% pa. Last year's rent was £1.85 million.

Due to its location, the site is particularly suitable for the existing tenant and if the existing tenant were not in place then it would be difficult to imagine re-letting the land for the same purpose. The alternative use would be as a general storage for which it would be worth around £25,000 per acre or £1.025m pa, or as building land for industrial units for which it could be sold for around £230,000 per acre or £9.43 million.

The total outlay will be £29 million plus stamp duty at 4%, agent's fees at 0.35%, legal fees at 0.25% and VAT on fees at 17.5% (now 20%) which comes to about £32.2 million.

## The advice

Consultant surveyors were employed to advise on the purchase. A summary of their recommendations is as follows.

First, the consultants outlined the 'running yield profile' of the asset by showing the relationship between the rent received in any year and the purchase price. This is set out as follows:

| Year | Rent | Running yield |
|------|------|---------------|
| 1 | £1.9627 | 6.10% |
| 2 | £2.0215 | 6.28% |
| 3 | £2.0822 | 6.47% |
| 4 | £2.1447 | 6.66% |
| 5 | £2.2090 | 6.86% |
| 6 | £2.2753 | 7.07% |
| 7 | £2.3435 | 7.28% |
| 8 | £2.4138 | 7.50% |
| 9 | £2.4862 | 7.72% |
| 10 | £2.5608 | 7.95% |
| 11 | £2.6377 | 8.19% |
| 12 | £2.7168 | 8.44% |
| 13 | £2.7983 | 8.69% |
| 14 | £2.8822 | 8.95% |
| 15 | £2.9687 | 9.22% |

The yield on reversion was expressed as 9.22%, with the reversion-ary yield beyond the 15-year point in 2020, dependent on whether the existing lessee renews the lease or the value reverts to industrial values.

If the tenant does not renew the lease, the yield on reversion is £1.025 m/£32.2 m = 3.2%. Note that this measure uses the current rental value and not a projection of the rental value of the land in year 15. This fixation on current values gives the property market participant an initial feel for the nature of the investment: 6.1% rising to 9.2% over 15 years falling back to 3.2% thereafter describes a particular pattern of risk and return which may be compared with a more straightforward, fully leased investment which yields 6% flat. The yield pattern described in the table above is misleading as it ignores the more likely yield on reversion.

The consultants then analysed the investment in terms of internal rate of return (IRR). A whole series of questions then arise, as follows:

- What holding period should be adopted?
- What assumptions should be used?

Fifteen years lends itself readily as the appropriate period, but five and 10 will also be of interest. An assumption then has to be made regarding the reversion.
- Will the tenant renew the lease?
- At what rent and under what terms will it be renewed?
- Will the tenant vacate? If the tenant vacates, the land may be sold as building land for industrial units for which it could be sold currently for around £230,000 per acre or £9.43 m. What will this value be in 15 years time?
- If the tenant renews then for what time period will the cash flow be projected?

For simplicity and in accordance with market practice, it may be best to assume a sale of the investment at year 15 following the re-letting. This comes back to comparable evidence.
- What initial or all risks yield would a buyer be likely to accept for the property if it has been let for an undefined period at a rent rising at 3% pa?

Assuming a re-letting at the year 15 rent with continued 3% uplifts the reversion is valued at an 'exit yield' of 6.75% (there was limited evidence to support this). The reversion can then be sold for £2.9687 million capitalised at 6.75% which gives a sale price of:

$$£\,2.9687\,/\,0.065 = £\,43.98\text{ million}$$

Assuming annual in arrears cash flows for simplicity, the cash flow then looks like this (in millions):

| Year | Capital | Rent | Total |
|------|---------|--------|---------|
| 0 | £32.20 | | −£32.20 |
| 1 | | £1.9627 | £1.9627 |
| 2 | | £2.0215 | £2.0215 |
| 3 | | £2.0822 | £2.0822 |
| 4 | | £2.1447 | £2.1447 |
| 5 | | £2.2090 | £2.2090 |

*Continued*

| Year | Capital | Rent | Total |
|------|---------|------|-------|
| 6 | | £2.2753 | £2.2753 |
| 7 | | £2.3435 | £2.3435 |
| 8 | | £2.4138 | £2.4138 |
| 9 | | £2.4862 | £2.4862 |
| 10 | | £2.5608 | £2.5608 |
| 11 | | £2.6377 | £2.6377 |
| 12 | | £2.7168 | £2.7168 |
| 13 | | £2.7983 | £2.7983 |
| 14 | | £2.8822 | £2.8822 |
| 15 | £43.98 | £2.9687 | £46.9487 |

Using the Excel IRR function, this is an IRR of 8.52%.

Assuming no re-letting and a sale of the land as building land, the consultants needed to make an assumption about the change in land values over the period. Land values have increased at something close to the rate of inflation in the long run, but in this case the consultant assumed that land would increase at rates varying from 2% to 6%. At 2% the cash flow changes in year 15 from £46.9487 to £15.6587 producing an IRR of 4.55%. The IRR rises to 5.19% with land value growth of 4% and 6.04% with land value growth of 6%.

Looking at the investment over five and 10-year holding periods, the difficulty is assessing the price at which the investment could be sold at the respective dates.

The consultants decided to split the cash flow at reversion into two parts. At year five the following year's cash flow of £32.28 million was split into the underlying rent of £25,000 per acre with rental growth of 4% applied and a remaining 'froth' or high risk element dependent upon continued occupation by the existing tenant. The rent components were:

£25,000 × 41 = £1.025 m

£1.025 × $(1.04)^5$ = £1.247 m. This is the Estimated Rental Value (ERV) component of rent.

The 'froth' component of rent is then £2.2090 m – £1.247 m = £0.963 m. Different exit yields were then applied to each component. 6.69% was applied to the ERV and 8% to the froth.

The exit value is then given by the following:

£1.247 m/0.0669 = £18.64 m
£0.963 m/0.08    = £12.04 m
Total            =£30.68 m

The five-year cash flow is terminated at year five with a year five value of £30.68 plus £2.2090, giving £32.89 and an IRR of 5.6%. A similar analysis was carried out by the consultants for a 10-year period using a slightly higher exit yield as there is increasing risk as the lease end approaches. The IRR on these 10-year assumptions came to 6.75%

## Modelling risk and return

The five, 10 and 15-year gilt redemption yields (in 2005) were all around 4.5%. So the question the consultants have to propose is 'Does this range of potential returns of, roughly, 4.5% to 8.5% provide adequate compensation for risk?'

- Have all the permutations been considered? It would be helpful to see an analysis of the likely return if the lease is renegotiated on reasonable terms (what are they likely to be?) and more variations regarding the planning position and alternative values.

- What are the relative probabilities of the higher and lower return outcomes? The permutations are many and a simulation may well be the best approach to this and similar problems.

    A further set of issues arise:

- Does the conventional valuation format using running yields add any value?

- What evidence does the valuer or consultant have for any choice of yields?

Explicit cash flow modelling is superior and the inadequacy of the traditional valuation approach is illustrated well in this illustration. However, the need to establish an exit rent and exit yield remains.

It is interesting to reflect on the market change from 2005 to 2011 and to ponder whether this level of analysis helps when the future is always going to be uncertain. 'What if' (?) is a question that must be asked; however the answer must be couched in terms of 'if this then that, if that then this' which may appear unhelpful but at least some of the possibilities will have been considered.

## Summary

- All valuation reports for all purposes should follow the minimum requirements of the RICS Valuation Standards unless an agreed report form is to be used.

- They should lead the reader to the same conclusions reached by the valuer and to do that they need to be unambiguous and honest.

- All facts need to be confirmed and all sources of information noted.

- The specific content may vary according to the purpose and the valuer needs to have special regard to the UK standards when reporting for secured lending, financial statements and other areas specifically covered by the UK standards.

- Investors require an increasing amount of information and will generally require something more than just an opinion of value. This may require the valuer to prepare various scenarios looking at possible re-sales after five, 10 and 15 years. These will not be market valuations but 'what if' calculations in respect of which the client might reasonably expect the property consultant to offer a professional view on the probability of different outcomes.

# Appendix C

## Illustrative Development Site Appraisal Report

The following report is based on the standard report produced by ARGUS Developer Version: 4.05.001 using the data from Example 10.1 on p203. Note that other software is available to undertake similar analysis.

Also for completion we have included the exhaustive list of assumptions that the user can change in the Argus programme.

Note: The land value calculated by the Argus system is not exactly the same as in the manual calculation due to rounding and minor variations within the assumptions made in the programme.

## Timescale and Assumptions

### Timescale (Duration in months)

Project commences Sep 2010

Phase 1

| Stage Name | Duration | Start Date | End Date |
|---|---|---|---|
| Phase Start | | Sep 2010 | |
| Construction | 24 | Sep 2010 | Aug 2012 |
| Phase End | | Sep 2012 | |
| **Phase Length** | **24** | | |
| **Project Length** | **25** | **(Includes Exit Period)** | |

*This first part shows the timescales and durations of the project. The development may be split into a number of phases which would be shown in this section. Phasing allows better manipulation and control of the cash flow. In this simple calculation there is no void period: this is unrealistic unless the development is fully pre-let.*

### Assumptions

### Expenditure

Professional Fees are based on Construction
Purchaser's Costs are based on Gross Capitalisation
Purchaser's Costs Deducted from Sale (Not added to Cost)
Sales Fees are based on Net Capitalisation
Sales Fees Added to Cost (Not deducted from Sale)

## Receipts

| | |
|---|---|
| Show tenant's true income stream | On |
| Offset income against development costs | Off |
| Rent payment cycle | Annual (Adv) |
| Apply rent payment cycle to all tenants | On |
| Renewal Void and Rent Free apply to first renewal only | Off |
| Initial Yield Valuation Method | Off |
| Default Capitalisation Yield | 0.0000% |
| Apply Default Capitalisation to All Tenants | Off |
| Default stage for Sale Date | Off |
| Align end of income stream to Sale Date | Off |
| Apply align end of income stream to all tenants | On |
| When the Capital Value is modified in the cash flow | Recalculate the Yield |
| Valuation Tables are | Annually in Arrears |
| Rent Free method | Defer start of Tenant's Rent |

## Finance

| | |
|---|---|
| Financing Method | Basic (Interest Sets) |
| Interest Compounding Period | Annual |
| Interest Charging Period | Annual |
| Nominal rates of interest used | |
| Calculate interest on Payments/Receipts in final period | Off |
| Include interest and Finance Fees in IRR Calculations | Off |
| Automatic Inter-account transfers | Off |
| Manual Finance Rate for Profit Erosion | Off |

## Calculation

| | |
|---|---|
| Site Payments | In Advance |
| Other Payments | In Advance |
| Negative Land | In Arrears |
| Receipts | In Advance |
| Initial IRR Guess Rate | 8.00% |
| Minimum IRR | −100% |
| Maximum IRR | 99999% |
| Manual Discount Rate | Off |
| IRR Tolerance | 0.001000 |

## Assumptions

| | |
|---|---|
| Letting and Rent Review Fees are calculated on | Net of Deductions |
| Development Yield and Rent Cover are calculated on | Rent at Sale Date(s) |
| Include Tenants with no Capital Value | On |
| Include Turnover Rent | Off |

| | |
|---|---|
| Net of Non-Recoverable costs | On |
| Net of Ground Rent deductions | On |
| Net of Rent Additions/Costs | On |

## Value Added Tax

| | |
|---|---|
| Global VAT Rate | 0.00% |
| Global Recovery Rate | 0.00% |
| Recovery Cycle every | 2 months |
| 1st Recovery Month | 2 (Oct 2010) |
| VAT Calculations in Cash Flow | On |
| GST Margin Calculations in Cash Flow | Off |

## Residual

| | |
|---|---|
| Land Cost Mode | Residualised Land Value |
| Multi-Phasing | Separate Land Residual for each phase |
| Target Type | Profit on GDV |

| Phase Number | Target Value | Locked Value | Treat Neg Land as Revenue |
|---|---|---|---|
| Phase 1 | 15.00% | No | No |

## Distribution

| | |
|---|---|
| Construction Payments are paid on | S-Curve |
| Sales Receipts are paid on | Single curve |
| Sales Deposits are paid on | Monthly curve |

## Interest Sets

### Site

| Debit Rate | Credit Rate | Months | Start Date |
|---|---|---|---|
| 15.000% | 0.000% | 24 | Sep 2010 |
| 0.000% | 0.000% | Perpetuity | Sep 2012 |

## Construction

| Debit Rate | Credit Rate | Months | Start Date |
|---|---|---|---|
| 15.000% | 0.000% | 24 | Sep 2010 |
| 0.000% | 0.000% | Perpetuity | Sep 2012 |

## Inflation and Growth

### Growth Sets

### Growth Set 1

Inflation/Growth for this set is calculated in arrears
This set is not stepped

| Rate | Months | Start Date |
|---|---|---|
| 0.000% | Perpetuity | Sep 2010 |

## Inflation Sets

### Inflation Set 1

Inflation/Growth for this set is calculated in arrears
This set is not stepped

| Rate | Months | Start Date |
|---|---|---|
| 0.000% | Perpetuity | Sep 2010 |

*This exhaustive set of assumptions gives the valuer many options and the flexibility to ensure the system mirrors the individual circumstances of the development project. Great care however must be exercised in ensuring that the assumptions, especially the default settings, are what are expected by the client.*

## APPRAISAL SUMMARY

### Summary Appraisal for Phase 1

**REVENUE**

| Rental Area Summary | m² | Rate m² | Gross MRV | |
|---|---|---|---|---|
| Office | 6,000.00 | £120.00 | 720,000 | |

**Investment Valuation Office**

| Current Rent | 720,000 | YP @ | 7.0000% | 14.2857 |
|---|---|---|---|---|
| **NET REALISATION** | | | | **10,285,714** |

**OUTLAY**
**ACQUISITION COSTS**

| | | | | |
|---|---|---|---|---|
| Residualised Price | | | 3,533,726 | |
| Stamp Duty | | 1.00% | 35,337 | |
| Agent Fee | | 1.00% | 35,337 | |
| Legal Fee | | 1.00% | 35,337 | |
| | | | | 3,639,738 |

**CONSTRUCTION COSTS**

| Construction | m² | Rate m² | Cost | |
|---|---|---|---|---|
| Office | 7,000.00 | £400.00 | 2,800,000 | **2,800,000** |

**PROFESSIONAL FEES**

| | | | | |
|---|---|---|---|---|
| Architect | | 9.00% | 252,000 | |
| Quantity Surveyor | | 3.50% | 98,000 | |
| | | | | 350,000 |

**MARKETING & LETTING**

| | | | | |
|---|---|---|---|---|
| Marketing | | | 50,000 | |
| Letting Agent Fee | | 20.00% | 144,000 | |
| | | | | 194,000 |

## DISPOSAL FEES

| Sales Agent Fee | 1.00% | 102,857 | |
|---|---|---|---|
| | | | 102,857 |

## FINANCE

Multiple Finance Rates Used (See Assumptions)
Debit Rates varied throughout the Cash Flow

| Total Finance Cost | 1,656,262 |
|---|---|
| **TOTAL COSTS** | **8,742,857** |

## PROFIT

| | **1,542,857** |
|---|---|

### Performance Measures

| Profit on Cost% | 17.65% |
|---|---|
| Profit on GDV% | 15.00% |
| Profit on NDV% | 15.00% |
| Development Yield% (on Rent) | 8.24% |
| Equivalent Yield% (Nominal) | 7.00% |
| Equivalent Yield% (True) | 7.32% |
| Gross Initial Yield% | 7.00% |
| Net Initial Yield% | 7.00% |
| | |
| IRR | 27.79% |
| Rent Cover | 2 yrs 2 mths |
| Profit Erosion (finance rate 15.000%) | 1 yr 2 mths |

*The system can be used for capital sales) e.g. residential development) as well as capitalised rental income as in this example, or a mixture of the two.*

*The expenditure summary shows additional items not considered in the simple manual calculation on page 307.*

*This page also shows some useful performance measures for the development including:*

*In this case the Profit on GDV has been stipulated by the user at 15% in order to calculate the residual site value. Where the system is used for calculating the developers profit he land value would be provided and the profit calculated by the system.*

*The profit erosion cover shows the period of time over which the profit would be completely eroded by continuing interest payments if the development remained un-let.*

*The rent cover is calculated by dividing the residual profit by the rental value. If the developer is guaranteeing rent as part of a funding arrangement a profit will still be made (decreasing over this period) providing the development is let within the period of years shown, in this case two years and two months.*

*The development yield is calculated by dividing the rental income by the total costs.*

*The other yields are explained in Chapter 12.*

## SENSITIVITY ANALYSIS

**Table of Residual Land Price**
Sensitivity Analysis for Phase 1

| Rent Rate | £ pm² | Construction Rate | | | | |
|---|---|---|---|---|---|---|
| | | −£20.00 m² | −£10.00 m² | £0.00 m² | +£10.00 m² | +£20.00 m² |
| | | £380.00 m² | £390.00 m² | £400.00 m² | £410.00 m² | £420.00 m² |
| −£5.00 m² | £115.00 m² | £3,407,180 | £3,340,514 | £3,273,848 | £3,207,181 | £3,140,515 |
| −£2.50 m² | £117.50 m² | £3,537,120 | £3,470,453 | £3,403,787 | £3,337,121 | £3,270,454 |
| £0.00 m² | £120.00 m² | £3,667,059 | £3,600,393 | £3,533,726 | £3,467,060 | £3,400,394 |
| +£2.50 m² | £122.50 m² | £3,796,998 | £3,730,332 | £3,663,665 −1,699,425 | £3,596,999 | £3,530,333 |
| +£5.00 m² | £125.00 m² | £3,926,937 | £3,860,271 | £3,793,605 | £3,726,938 | £3,660,272 |

## Sensitivity Analysis: Assumptions for Calculation

### Construction Rate

| Heading | Phase | Original Value |
|---------|-------|----------------|
| Office  | 1     | £400.00 pm$^2$ |

These fields varied in Fixed Steps of £10.00 pm$^2$

### Rent Rate

| Heading | Phase | Original Value |
|---------|-------|----------------|
| Office  | 1     | £120.00 pm$^2$ |

These fields varied in Fixed Steps of £2.50 pm$^2$

*The system can be used to generate a sensitivity analysis matrix.*

*In this case the matrix shows the effect on the residual land value with changes in two variables: the construction cost rate and the rental rate. The rental rate changes in steps of £2.50 and the building costs in steps of £10.*

*The 'worst' case scenario is therefore in the top right-hand corner of the matrix and the 'best' case scenario in the bottom left hand corner. Sensitivity here is not that significant, although when analysing yields however very small changes in the yield may result in significant changes in the residual value or profit of a scheme.*

*The original residual value (£3,553,726); calculated with the unadjusted values for rent and construction is shown in the centre of the table.*

*When using the system for appraisals, the matrix would show the amount of profit in £ (or other currency) and as a percentage of either Cost, GDV or NDV as selected by the user. The sensitivity can also produce a three dimensional matrix by using five layers or windows which are printed separately.*

Report produced by
ARGUS Developer Version: 4.05.001

# Appendix D

## Solutions to Questions Set in the Text

### Chapters 2 and 3

1. £975.33 (A £1)
2. £663.429 (present value (PV) £1)
3. £3,553.20 (Annuity £1× £40,000 or ÷ PV £1 pa)
4. £21,321.44 (A £1 pa)
5. £97.38 (annual sinking fund (ASF))
6. (a) Lump sum or ASF
   (b) £8,049.25 or £1,424.60 pa
   (c) £25,000 × A£1 at 12% = £25,000 × 3.1058 = £77,645 × PV £1 at 12% = £25,000 ∴ £25,000 or £77,645 × ASF 0.05698 = £4,424.21
7. (a) £2,763 (A £1 pa)
   (b) £5,764.75
   (c) £8,085.36
8. £465,292 (£10,000 × PV £1 pa at 8% + £50,000 × PV £1 pa in perp at 8% × PV £1 in 10 years at 8%)
9. 13%

### Chapter 4

Q1. (75,000/1,250,000) × 100 = 6% [All risks yield (ARY) or capitalisation rate (cap rate) or initial yield in this instance. Note: (1,250,000/75,000)=16.667 (PV £1 pa in perp) and (100/16.667) = 6%

Q2. This is an equivalent yield analysis which can be solved using Goal Seek in Excel or trial and error. The trial and error format is shown here.

| | | | |
|---|---|---|---|
| Term income | £30,000 | | |
| PV £1 pa for 2 years at x% | (1.8080) | (£54,240) | |
| Reversion to MR | £40,000 | | |
| PV £1 pa in perp at x% | | | |
| X PV £1 in 2 years at x% | (12.477) | (£499,080) | (£533,000) |
| | | | £533,000 |

Using trial and error all figures in parentheses are unknown and so the analyst has to try a solution at say 6% which will produce value above £533,000, while at 8% the value will be below £533,000; this repetitive process can be quite lengthy for a student under assessment conditions; hence in practice the use of Goal Seek or alternative electronic solutions. Here, substituting 7% for x% will produce the PV and PV £1 pa figures shown in the brackets.

Analysis must be consistent, here the analysis has been undertaken without deduction of management fees from the net rents, but provided one values as one analyses then analysis with or without such a deduction when let on full repairing and insurance (FRI) terms will be acceptable.

# Chapter 5

## Q1 Implied growth rate calculation

ARY (k) = 6%
Gilts = 8%
Target rate ($e$) = 8% + 1% = 9%
Rent review ($t$) = 5 years
Substituting in: $k = e - (ASF \times P)$
$0.05 = 0.09 - (0.16709 \times P)$
$0.16709P = 0.09 - 0.06$
$0.16709P = 0.03$
$P = 0.03/0.16709$
$P = 0.18$ (18% over 5 years)
Therefore annual growth is:
$(1+g)^5 = 1.18$
$(1+g) = \sqrt[5]{1.18}$
Or $(1+g) = 1.18^{0.20}$
$(1+g) = 1.03365$
$g = 1.03365 - 1$
$g = 0.03365 \times 100$
$g = 3.365\%$ per year

## Q2 Value a freehold warehouse

Step 1: Find the ARY
$(£42,500/£472,000) \times 100 = 9.0\%$
Step 2: Find market rent from comparable
Given as £42,500
Step 3: Equivalent yield valuation
Assume no market variation from ARY

| | | |
|---|---:|---:|
| Term income | £30,000 | |
| PV £1 pa for 2 years at 9% | 1.7591 | £52,773 |
| Reversion to MR | £42,500 | |
| PV £1 pa in perp defd 2 years at 9% | 9.3520 | £397,460 |
| | | £450,233 |

Step 4: Short cut DCF
Calculate implied growth rate where:
$K = 0.09$
$e = 0.12$
$t = 5$ year rent reviews
and $K = e - (ASF \times P)$

| | |
|---|---|
| 0.09 | $= 0.12 - (0.15741P)$ |
| 0.15741P | $= 0.12 - 0.09$ |
| 0.15741P | $= 0.03$ |
| P | $= 0.03/0.15741$ |
| P | $= 0.19058$ say 0.19 |

Therefore implied annual growth is:

| | |
|---|---|
| $(1+ g)^5$ | $= 0.19$ (A£1 formula is $(1+i)n$ is $(1+$ interest to produce £1.19 in 5 years) |
| $(1+ g)$ | $= \sqrt[5]{1.19}$ |
| $(1+ g)$ | $= 1.0354$ |
| g | $= 0.0354 \times 100$ |
| g | $= 3.54\%$ |

| | | | |
|---|---|---|---|
| Term income | | £30,000 | |
| PV £1 pa at 12% | | 1.6901 | £50,703 |
| Reversion to MR | | £42,500 | |
| A £1 in 2 years at 3.54% | | 1.072 | |
| | | £45,562 | |
| PV £1 pa in perp at 9% | 11.111 | | |
| PV £1 in 2 years at 12% | 0.79719 | 8.8575 | £403,571 |
| | | | £454,274 |

Step 5: Real Value Approach
Assess IRFY using: $\dfrac{e - g}{1 + g}$

$$\frac{0.12 - 0.0354}{1.0354} = \frac{0.0846}{1.0354} = 0.0817 = 8.17\%$$

| | | | |
|---|---|---|---|
| Term income | | £30,000 | |
| PV £1 pa at 12% | | 1.6901 | £50,703 |
| Reversion to MR | | £42,500 | |
| PV £1 pa in perp at 9% | 11.111 | | |
| PV £1 in 2 years at 8.17% | 0.8546 | 9.4959 | 403,578 |
| | | | £454,281 |

The marginal differences in total are due to rounding of some numbers.

In this question the variation in opinion of value is small because the reversion is short i.e. in 2 years time; the implied rate of rental growth is low at 3.54% and the current term income is 70% of the market rent. The variation between DCF based value opinions and income capitalisations increases when the reversion is distant i.e. over say 20 years; rental growth is substantial say above 5% a year and the current rent is very low compared to current market rent say below 50%.

# Chapter 8

## Q1 Premium

Freehold ARY = 5%

| | |
|---|---|
| MR | £75,000 |
| Proposed rent | £50,000 |
| Loss of rent to freeholder | £25,000 |
| PV £1 pa for 5 years at 5% | 3.502 |
| Premium needed | £87,547 |

In theory a tenant's view could be found using a higher leasehold yield or dual rate but this will instantly produce a lower PV figure or multiplier and therefore a non-negotiable position.

## Q2 Surrender and renewal

**(a)**
**(i)** Freeholder's present interest

| | | |
|---|---|---|
| Current rent | £10,000 | |
| PV £1 pa for 10 years at 9% | 6.4177 | £64,177 |
| Reversion to MR | £100,000 | |
| PV £1 pa perp defrd 10 years at 9% | 4.6935 | £469,350 |
| | | £533,527 |

The ARY has been adjusted from the market evidence by 1% to reflect the 10-year term.

**(ii)** Freeholder's proposed interest

| | | |
|---|---|---|
| Rent to be reserved | £x | |
| PV £1 pa for 5 years at 8% | 3.9927 | £3.9927x |
| Reversion to MR | £100,000 | |
| PV £1 pa perp defd 5 years at 8% | 8.5073 | £850,730 |
| | | £850,730 + £3.9927x |

Let:

| Value of present interest | = Value of proposed interest |
|---|---|
| £533,527 | = £850,730 + £3.9927x |
| £533,527-£850,730 | = £3.9927x |
| −£317,203 | = £3.9927x |
| −£79,445 | = £x |

This surrender and renewal is sufficiently beneficial for the freeholder to pay £79,445 a year to the tenant for the next five years.

**(iii)Tenant's present interest**

| Market rent | £100,000 | | |
|---|---|---|---|
| Rent payable | | £10,000 | |
| Profit rent | £90,000 | | |
| PV £1pa for 10 years at 12% | | 5.6502 | £508,518 |

**(iv) Tenant's proposed interest**

| Market rent | £100,000 | | |
|---|---|---|---|
| Rent proposed | | £x | |
| Profit rent | £100,000−£x | | |
| PV £1 pa for 5 years at 12% | | 3.6048 | £360,480−£3.6048x |

| Let present interest | = proposed interest |
|---|---|
| £508,518 | = £360,480−£3.6048 |
| £508,518 − £360,480 | = −£3,6048 |
| £148,038 | = −£3.6048 |
| £148,038 ÷ 3.6048 | = −£x |
| £41,067 | = −£x |

Provided the 12% is supported by market evidence then the tenant will accept no less than £41,067, and the landlord will offer no more than £79,445 by way of an annual payment to the tenant. Readers need to be certain of their arithmetic rules about positive and negative signs and movement from one side of an equation to another. It is clear here that the tenant's profit rent has to get better as they are surrendering a 10-year profit rent for a five-year profit rent, for values to equate the profit rent for 5 years must be bigger.

**(b)Freeholder's present interest as in (i) above = £533,527**
**(v)Freeholder's proposed interest**

| Market rent | £100,000 | | |
|---|---|---|---|
| PV £1 pa in perpetuity at 8% | | 12.5 | £1,250,000 |
| Less reverse premium | | | £x |
| | | | £1,250,000−£x |

| Let present interest | = proposed interest |
|---|---|
| £533,527 | = £1,250,000−£x |
| £x | = £1,250,000−£533,527 |
| £x | = £716,473 |

**(vi)** Tenant's interest as in (iii) = £508,518
**(vii)** Tenant's proposed interest

| | |
|---|---|
| Market rent | £100,000 |
| Rent to be paid | £100,000 |
| Profit rent | £0 |

Tenant's proposed interest is going to be nought and therefore a sum of £508,518 will need to be paid by the landlord to the tenant and the landlord will be willing to pay up to £716,473 and so negotiations should arrive at a mutually acceptable figure.

This problem could be reset as a marriage value exercise. The freehold value in possession is £1,250,000, while the freehold subject to the lease is £533,527 and the leasehold is £508,518. The combined value is £1,042,045 and hence the marriage value or merger value of £207,955 exists (£1,250,000 – £1,042,045).

In the fifth edition, and someone is bound to do the comparison, dual rate adjusted for tax was used at 10% and 3% adjusted for tax at 40p in the £. Of course in a merger situation there would be no tax issue and no capital replacement issue and hence no justification for dual rate adjusted for tax. However, of interest is the fact that dual rate produced a lower leasehold value and a higher merger value. Hence our warning that it is not the maths which generates merger value but the market, so who is the market and what is the market value?

## Q3 Non standard rent review

The formula approach as set out in Chapter 8 produces a multiplier k to correct the known rent to the unknown rent, thus:

$$k = \frac{(1+r)^n - (1+g)^n}{(1+r)^n - 1} \times \frac{(1+r)^t - 1}{(1+r)^t - (1+g)^t}$$

Where:
$K$ = multiplier
$r$ = equated yield here of 10% = 0.10
$n$ = number of years to review in subject lease here = 3
$g$ = annual rental growth = 0.05 (this might have to be calculated using the implied rental growth rate formula
$t$ = number of years to review normally agreed which relates to the market rent evidence.
And substituting:

$$\frac{(1+0.13)^3 - (1+0.05)^3}{(1+0.10)^3 - 1} \times \frac{(1+0.10)^5 - 1}{(1+0.10)^5 - (1+0.05)^5}$$

$$= \frac{1.331 - 1.1576}{1.331 - 1} \times \frac{1.6105 - 1}{1.6105 - 1.2763} = \frac{0.1734}{0.331} \times \frac{0.6105}{0.3342}$$

$$\therefore \ £25,000 \times 0.9568 = £23,920$$

A landlord in a 5% a year rising market will be willing to accept either £23,920 with rent reviews every three years or £25,000 with rent reviews every five years. In theory the position is reversed in a falling market with landlords willing to accept less rent for the security of a longer term.

# Chapter 9

## Q1 Landlord and tenant issues

**(a)** Market rent improved

| | | |
|---|---|---|
| 4th floor | 250 m² × £200 | = £50,000 |
| Ground,1st, 2nd, 3rd floors | 400 m² × 4 × £200 | = £320,000 |
| MR | | £370,000 |

Section 34 market rent disregarding improvements

| | | |
|---|---|---|
| 1st, 2nd, 3rd floors | 400 m² × 3 × £150 | £180,000 |
| | 400 m² × £200 | £80,000 |
| | | £260,000 |

Probable S 34 rent is £260,000.

**(b)** The freehold interest with LPA 21 year rule

| | | | |
|---|---|---|---|
| Current net income | | £150,000 | |
| PV £1 pa for 3 years at 7% | | 2.6243 | £393,645 |
| Reversion to S 34 rent | | £260,000 | |
| PV £1pa for 15 years at 7% | 9.1079 | | |
| PV £1 for 3 years at 7% | 0.81630 | 7.4347 | £1,933,042 |
| Reversion in 18 years to MR | | £370,000 | |
| PV £1 pa in perp. defd 18 years at 7% | | 4.2266 | £1,563,842 |
| | | | £3,890,529 |

The courts might allow for a rent review after 11 years to market rent, i.e. immediately after the 21 years has run out, but this would appear to contradict the Landlord and Tenant Act 1954 and LPA expectation. For a valuer to assume in a valuation something which is relatively uncertain could be held to be being too optimistic and possibly negligent; hence the period of 15 years has been used. Students please note that this just happens to be the maximum lease length a court could grant, but that is not the reason for adopting 15 years here. It can be helpful to plot everything on a line diagram or as a logical time flow. There will be rent reviews during the 15 years but these in conventional valuation terms will be to today's market rent and again it is a reasonable assumption that a competent tenant solicitor will insist on section 34 style disregards being inserted in the rent review clause.

**(c)** Tenant's leasehold interest

| | | | |
|---|---|---|---|
| MR | | £370,000 | |
| Rent reserved | | £150,000 | |
| Profit rent | | £220,000 | |
| PV £1 pa for 3 years at 14% | | 2.3216 | £510,752 |
| S 34 reversion | | | |
| MR | | £370,000 | |
| Rent reserved (S34) | | £260,000 | |
| Profit rent | | £110,000 | |
| PV £1 pa for 15 years at 14% | 6.1422 | | |
| PV £1 pa for 3 years at 14% | 0.67497 | 4.1458 | £456,038 |
| | | | £966,790 |

If dual rate is used then there will be a variable profit rent sinking fund problem to solve.

If the tenant acquires the freehold then the interests will be merged and would have a market value of:
Freehold in possession

| | | |
|---|---|---|
| MR | £370,000 | |
| PV £1 pa in perpetuity at 7% | 14.2857 | £5,285,709 |

A professionally advised freeholder will be aware of the potential merger value and will seek a price for the freehold in excess of the market value of £3,890,529. Generally, 50% of the market value will need to be offered on top of the merger value of the freehold interest. So, £5,285,709 less the freehold value of £3,890,529 less the leasehold value of £966,790 leaves a merger value of £428,390 split two ways is an extra £214,195 for the freeholder.

**(c)** Surrender for new 20 year lease with a rent review to MR after 5 years.
Landlord's freehold interest as before £890,529
Landlord's proposed interest

| | | |
|---|---|---|
| Let rent for first 5 years be | £x | |
| PV £ pa for 5 years at 7% | 4.1002 | £4.1002x |
| Reversion to MR | £370,000 | |
| PV £ pa in perp defd 5 years at 7% | 10.1855 | £3,768,635 |
| | | £3,768,635 + £4.1002 x |

| | |
|---|---|
| Present | = proposed |
| £3,890,529 | = £3,768,635 + £4.1002x |
| £3,890,529−£3,768,635 | = £4.1002x |
| £121,894 | = £4.1002x |
| £29,722 | = £x |

Tenant's view
Present interest as before at £966,790
Tenant's proposed interest

| | |
|---|---|
| MR | £370,000 |
| Rent to be reserved | £x |
| Profit rent | £370,000-£x |
| PV £1 pa for 5 years at 14% | 3.4331 |
| | £1,270,247 − £3.4331x |

| Present | = proposed |
|---|---|
| £966,790 | = £1,270,247 − £3.4331x |
| £3.4331x | = £1,270,247 − £966,790 |
| £3.4331x | = £303,457 |
| £x | = £303,456 ÷ 3.4331 |
| £x | = £88,392 |

This gives ample room for negotiations between the landlord's £29,722 and the tenant's £88,392. As this is a surrender and renewal, the tenant loses all their rights under the 1954 Act on surrender. So after five years the rent will rise to market rent. Some may argue that 14% is not high enough, and that any attempt to sell an occupation lease such as this is going to be problematic, and 20% might be more appropriate. Again, do not let the maths dictate the negotiations, it is market value before and after which is critical, and evidence of similar leasehold transactions is going to be very weak.

**(d)** Repossession can only be obtained on the following grounds:
(i) breach of repairing covenants
(ii) persistent delay or failure to pay rent
(iii) other substantial breach of covenants
(iv) availability of suitable alternative accommodation
(v) uneconomic sub-letting
(vi) demolition or substantial reconstruction
(vii) required by the owner for his own occupation (five-year rule applies).

The amount of compensation will be zero for (i),(ii),(iii) and (iv) and will be the rateable value in the other cases or twice the rateable value if the tenant or predecessor in title have been in occupation for 14 years or longer.

In addition, the tenant will be entitled to compensation for the improvements under the Landlord and Tenant Act 1927, being the lesser of the cost at the time of termination or net addition to value.
Here:

(a) net addition to value is

| | | |
|---|---|---|
| Improved MR | £370,000 | |
| Unimproved MR | £260,000 | |
| Increase in MR | £110,000 | |
| PV £1 pa in perpetuity at 7% | 14.2857 | £1,571,427 |

(b) Cost

| | | |
|---|---|---|
| 7 years ago the works cost | | £300,000 |
| Allow for cost inflation as per construction cost index | | |
| At say 5% per year X A£1 pa for 7 years at 5% | 1.2763 | £382,890 |

It is clear that the cost is the lesser sum and that compensation for improvements will be £382,890 adjusted for any dilapidations, plus compensation for loss of security of tenure at twice the rateable value of £350,000 is £700,000 giving a total of £1,050,000.

# Chapter 10

## Q1 Residual site value

Gross development value (GDV)
5000 m² (gross) × 855 = 4,250 m² net

| | | |
|---|---|---|
| 4,150 m² at £200 m² pa | £850,000 | |
| PV £1 pa in perp at 7% | 14.29 | £12,146,500 |
| Less purchaser's costs and stamp duty at 5.75% (÷1.0575) | | £325,089 |
| | | £11,486,052 |

Less costs

| | | |
|---|---|---|
| (a) Building 5,000 m² at £400 m² | £2,000,000 | |
| (b) Fees at 12.5% | £250,000 | |
| Total | £2,250,000 | |
| (c) Finance 14% for 1 year on 50% of total cost | £157,500 | |
| (d) Legal fees 1% and agent's fees at 1% on sale and promotion | £276,428 | |
| (e) Profit at 15% of GDV | £1,821,975 | £4,505,903 |
| NDV | | £6,980,149 |

| | |
|---|---|
| Let site value | = £1x |
| Fees on acquisition of site and stamp duty on site purchase | = 4% + 1.75% = 5.75% = £0.0575x |
| Total debt after 1 year with interest at 14% | = £1.0575x ×(1.14) |
| £6,980,149 | = £1.2055x |
| Site value | = £6,980,149 ÷ 1.2055 = £5,790,252 |
| Say £5,790,000 | |

This short residual calculation would provide the basis for an initial assessment. It could be compared on a per hectare/acre basis and also as a percentage of GDV. Provided in ball park figures it looks to be a comparable figure - then a more thorough calculation with sensitivity analysis and a cash flow can be undertaken using a software package (see Appendix C).

# Chapter 12

## Q1 Investment questions

**(a)** IRR = internal rate of return; that is the discount rate that makes the net present value (NPV) of a project equal to zero.

**(b)** The IRR of an investment before allowing for the incidence of taxation.

**(c)** The IRR of a property investment after adjustment for acquisition costs, outgoings but not taxation, taking account of current income and reversionary incomes expressed in current value terms.

**(d)** The IRR of a property investment after adjustment for acquisition costs, outgoings but not taxation, taking account of current income and reversionary incomes expressed in future value terms.

## Q2 Property question

**(a)** £1,000,000 + stamp duty at 4% = £1,040,000 + other fees of say 1.75% = £1,058,200.

**(b)** (£60,000 ÷ £1,058,200) ×100 = 5.67%.

**(c)** (£110,000 ÷ £1,058,200) × 100 = 10.40%.

**(d)** Trial and error or Goal Seek in Excel if analysis package set up on Excel.

| | | |
|---|---|---|
| Term rent | £60,000 | |
| PV £1 pa for 4 yrs at try 9.5% | 3.2045 | £192,270 |
| Reversion to | £110,000 | |
| PV £1 pa in perp defd 4 yrs at 9.5% | 7.32183 | £805,401 |
| Term + reversion | | £997,671 |

At a trial rate of 9.5%, the total is just under the £1,000,000 figure and so 9.5% is fractionally too high. For reporting purposes Goal Seek would probably be used which would give 9.48% (9.48036688%).

**(e)** Rental growth at 5% therefore rent on review is £110,000 × (1 + .05)$^4$ = £133,705. Value in 4 years time at 9.5% will be £133,705 × $\left(\dfrac{1}{9.5}\right)$ = £133,705 × 10.5263 = £1,407,421.

Therefore, possible cash flow adjusted for growth would be:

| | |
|---|---|
| 0 | − £1,040,000 |
| 1 | + £60,000 |
| 2 | + £60,000 |
| 3 | + £60,000 |
| 4 | + £60,000 |
| | + £1,407,421 |

Again, trial and error could be used to find the IRR of this cash flow but it is assumed that readers are now using Excel and will be able to solve this with the IRR function or Goal Seek and the IRR should be 13.048%.

**(f)** The requirement is for an IRR or target rate of 12%; this could be solved using a short cut discounted cash flow and solving to find the necessary rent in four years' time.

| | | |
|---|---|---|
| Term rent | £60,000 | |
| PV £1 pa for 4 yrs at 12% | <u>3.0373</u> | £182,238 |
| | | |
| Reversion to | £x | |
| PV £1 pa in perp at 9.5% x PV £1 | <u>6.6689</u> | <u>£6.6689x</u> |
| in 4 yrs at 12% | | |
| | | £182,238 + £6.689x |

But £182,238 + £6.689x − £1,040,000 must = NPV of zero ∴

£6.6896x = £1,040,000 − £182,238

£x = £857,762 ÷ 6.689

£x = £128,234.71

The rent in year four will need to be £128,234.

**(i)** £18,234.71

**(ii)** £110,000 x A £1 for 4 years at x% = £128,234.71

A= £128,234.71 ÷ £110,000 = 1.16577

(1.16577−1)× 100 = 16.577% over 4 years

If $(1+i)^4 = 1.6577$

Then $t = \sqrt[4]{1.6577}$ minus 1 = 1.03909 − 1 = 0.03909 × 100 = 3.909%

Say 3.9%

# Appendix A

## Q1 Dual rate

| | |
|---|---|
| Profit rent | £20,000 |
| PV £1 pa 4 years dual rate at 8%/3% tax at 40% | <u>2.0904</u> |
| | £41,808 |
| Less 8% gross return | <u>£3,344.64</u> |
| Balance for SF and tax on SF contribution | £16,655.36 |
| Less tax at 40% | <u>£6,662.14</u> |
| SF | £9,994.22 |
| X A £1 pa for 4 years at 3% | <u>4.1836</u> |
| Sinking fund accumulation is | £41,811.82 |

There is a small rounding error in this calculation.

## Q2 Dual rate and DCF

Conventional valuation of 60 year unexpired lease using dual rate

| | | | |
|---|---|---|---|
| Current rent | | £2,500,000 | |
| Head rent | | £50,000 | |
| Profit rent | | £2,450,00 | |
| YP dual rate 3 years at 8%/3% tax 40% | | 1.6149 | £3,956,505 |
| Reversion to | | £3,250,000 | |
| Head rent | | £50,000 | |
| Profit rent | | £3,200,000 | |
| YP dual rate 7 years | | 2.6682 | £8,538,308 |
| at 8%/3% tax 40% | 3.3612 | | |
| XPV £1 in 3 years at 8% | 0.79383 | | |
| Reversion to | | £3,250,000 | |
| Head rent 5% | | £162,500 | |
| Profit rent | | £3,087,500 | |
| YP dual rate 50 years | | 4.8872 | £15,089,261 |
| at 8%/3% tax 40% | 10.5512 | | |
| XPV £1 in 10 years at 8% | 0.46319 | | |
| Total | | | £27,584,074 |

However, note:

- Variable profit rents make an over provision for sinking fund payments, only one fund is needed over 60 years.
- So need to rework using sinking fund method as set out in Appendix A.
- This approach fails to reflect the real growth prospects of this type of investment.
- The head rent in year 10 and the market rent in year 10 are both based on today's estimate of market rent. Here lies one of the problems.

## DCF valuation

Step 1: Calculate the implied rate of rental growth using the rental growth formula. This produces a figure of 3.42% pa given a freehold ARY of 7% and an equated yield of 10% with rent reviews every five years.
Step 2: Create a spreadsheet using Excel.
Step 3: Complete the DCF.

| Period (end of year) | Shop Rents | Head Rent | Profit Rent at 40p | Net of Tax at 7.2% | PV £1 pa at 7.2% | PV £1 | PV |
|---|---|---|---|---|---|---|---|
| 1 | 2,500,000 | 50,000 | 2,450,000 | 1,470,000 | | 0.9328 | 1,371,216 |
| 2 | 2,500,000 | 50,000 | 2,450,000 | 1,470,000 | | 0.8702 | 1,279,194 |
| 3 | 2,500,000 | 50,000 | 2,450,000 | 1,470,000 | | 0.8117 | 1,193,199 |
| 4 | 3,594,983 | 50,000 | 3,544,983 | 2,126,989 | | 0.7572 | 1,610,556 |
| 5 | 3,594,983 | 50,000 | 3,544,983 | 2,126,989 | | 0.7063 | 1,502,292 |
| 6 | 3,594,983 | 50,000 | 3,544,983 | 2,126,989 | | 0.6589 | 1,401,473 |
| 7 | 3,594,983 | 50,000 | 3,544,983 | 2,126,989 | | 0.6146 | 1,307,247 |
| 8 | 3,594,983 | 50,000 | 3,544,983 | 2,126,989 | | 0.5734 | 1,219,615 |
| 9 | 4,253,225 | 50,000 | 4,203,225 | 2,521,935 | | 0.5348 | 1,348,731 |
| 10 | 4,253,225 | 50,000 | 4,203,225 | 2,521,935 | | 0.4989 | 1,258,193 |
| 11 | 4,253,225 | 212,661 | 4,040,564 | 2,424,338 | | 0.4654 | 1,128,287 |
| 12 | 4,253,225 | 212,661 | 4,040,564 | 2,424,338 | | 0.4342 | 1,052,647 |
| 13 | 4,253,225 | 212,661 | 4,040,564 | 2,424,338 | | 0.4050 | 981,857 |
| 14-18 | 5,031,991 | 212,661 | 4,819,330 | 2,891,598 | 4.078 | 0.4050 | 4,775,734 |
| 19-23 | 5,953,348 | 212,661 | 5,740,687 | 3,444,412 | 4.078 | 0.2860 | 4,017,245 |
| 24-28 | 7,043,406 | 212,661 | 6,830,745 | 4,098,447 | 4.078 | 0.2021 | 3,377,791 |
| 29-33 | 8,333,054 | 212,661 | 8,120,393 | 4,872,235 | 4.078 | 0.1427 | 2,835,302 |
| 34-38 | 9,858,837 | 212,661 | 9,646,176 | 5,787,705 | 4.078 | 0.1008 | 2,379,108 |
| 39-43 | 11,663,990 | 212,661 | 11,451,329 | 6,870,797 | 4.078 | 0.0712 | 1,994,961 |
| 44-48 | 13,799,966 | 212,661 | 13,587,305 | 8,152,383 | 4.078 | 0.0503 | 1672244 |
| 49-53 | 16,326,739 | 212,661 | 16,114,078 | 9,668,446 | 4.078 | 0.0355 | 1,399,695 |
| 54-58 | 19,316,165 | 212,661 | 19,103,504 | 11,462,102 | 4.078 | 0.0251 | 1,173,235 |
| 59-60 | 22,852,955 | 212,661 | 22,640,294 | 13,584,176 | 1.803 | 0.0177 | 433,513 |

The total of the final column is £40,713,331 which suggests that the conventional dual rate valuation has considerably undervalued this investment.

Note:

1. 12% gross less tax at 40% is 7.25 net.
2. £3,250,000 the market rent rises at 3.42% to £3,594,983 in three years time; then to £4,253,225; then to £5,031,991 and finally for the last five years is £22,852,995.
3. The affect of this is to change the head rent from the conventional assumption but this still has very little impact on the value.
4. This DCF may not be the market value. It may be an assessment of investment worth. However, unless valuers look and see 'what if?' they may under or over value from the point of view of potential buyers. In the case of significant leasehold investments, the valuation should be gross as the main buyers are gross funds or the other party, i.e. typically a freeholder buying back a lease.

# Further reading
## and Bibliography

The previous editions of this book contained an extensive bibliography of the many journal articles and books which had been referred to by us in preparing the first edition. This shortened reading list and bibliography covers those publications noted in the text and a few that readers may want to look at for historical reasons. Many of the ideas covered in earlier articles in particular have now been subsumed in current practice and theory.

Baum, A. (2009) *Commercial Real Estate Investment, A strategic Approach*, 2nd ed, London, EG Books.

Baum, A. and Crosby, N. (2007) *Property Investment Appraisal*, 3rd ed, Oxford, Blackwell.

Baum, A., Sams, G., Ellis, E., Hampson, C. and Stevens, D. (2007) *Statutory Valuations*, London, EG Books.

Bowcock, P. (1978) *Property Valuation Tables*, London, Macmillan.

Bowcock, P. (1983) *The Valuation of Varying Incomes,* Journal of Property Valuation and Investment, Vol 1. No 4.

Bowcock, P. and Bayfield, N. (2003/2004) *Excel for Surveyors and advanced Excel for Surveyors*, London, Estates Gazette.

Brooker v Unique Pub Properties Ltd (2001); & Brooker v Unique Properties Ltd (2009).

Byrne, P.J. and Mackmin, D.H. (1975) *The Investment method; an objective approach*, Estates Gazette 234:29.

Cairncross A. (1982) *Introduction to Economics,* London, Butterworth Heinemann Ltd.

Crehan v Inntrepreneur Pub Co (CPC) & Anor (2003) High Court EWHC 1510(ch)/(2004) EWCA civ 637/2006 UKHL 38

Crosby, N. (1983) *The Investment Method of Valuation; a Real Value Approach*, Journal of Valuation no.1, 341–50, 2:48–59.

Crosby, N. and Godchild, R. (1992) *Reversionary Freeholds; Problems with Over-Renting*, Journal of Property Valuation and Investment, 11:67–81.

Damodaran, A. (2001) *The Dark Side of Valuation,* New Jersey, Prentice Hall.

Davidson, A.W. (2002) *Parry's Valuation and Investment Tables*, 12th ed, London, Estates Gazette.

Day, H. and Kelton, R. (2007) *The Valuation of Licensed Premises*, Journal of Property Investment & Finance, Vol 25. No 3.

Dubbin, N. and Sayce, S. (1991) *Property Portfolio Management*, London, Routledge.

Emeny, R. and Wilks, H.M. (1982) *Principles and Practice of Rating Valuation*, London, Estates Gazette.

Fraser, W.D. (1984) *Principles of Property Investment and Pricing*, London, Macmillan; and 2nd edition (1993).

Greenwell, W. (1976) *A Call for New Valuation Methods*, Estates Gazette 238: 481.

Harker, N., Nanthakumaran, N. and Rogers, S. (1988) *Double Sinking Fund Correction Methods*, University of Aberdeen discussion paper.

IVSC (2005) *International Valuation Standards*, International Valuation Standards Committee.

Langdon, D. (Compilers) (2011) *Spon's Architects' and Builders' Price Book*, London, Davis Langdon.

Mackmin, D. (2008) *Valuation and Sale of Residential Property*, London, EG Books.

Mackmin, D. (2008) *Dual Rate is Defunct/A Review of Dual Rate Valuation, its History and its Irrelevance in Today's UK Leasehold Market,* Journal of Property Investment Finance 26,1.

Shapiro, E., Davies, K. and Mackmin, D. (2009) *Modern Methods of Valuation*, 10th ed, London, EG Books.

Marshall, P. (1979) *Donaldson's Investment Tables*, 2nd ed, London, Donaldsons.

Marshall, H. and Williamson H. (eds) (1996) *Law & Valuation of Leisure Property*, London, Estates Gazette.

Martin, S.A. (editor) (1987) *Real Estate Appraising in Canada,* Winnipeg, Canada, Appraisal Institute of Canada.

Musto, N.E., Eve, H.B. and Anstey, B. (1938) *Complete Valuation Practice,* London, Estates Gazette.

RICS, *Discounted Cash Flow for Commercial Property Investments,* RICS guidance note (GN7) (rics.org/standards), RICS Valuation Professional Group, RICS, London.

RICS Practice Standards, UK (GN67/2010) *The Capital and Rental Valuation of Public Houses, Bars, Restaurants and Nightclubs in England and Wales,* 1st ed, guidance note.

RICS Valuation Information Paper No 2, *The Capital and Rental Valuation of Public Houses, Bars, Restaurants and Nightclubs in England and Wales* (3rd ed *in preparation*).

RICS Valuation Information Paper No 3 (2003) *The Capital and Rental Valuation of Petrol Filling Stations in England, Wales and Scotland.*

RICS Valuation Information Paper No 6 (2004) *The Capital and Rental Valuation of Hotels in the UK.*

RICS Valuation Information Paper No 11 (2007) *Valuation and Appraisal of Private Care Home Properties in England, Wales and Scotland.*

RICS Valuation Information Paper No 12 (2008) *The Valuation of Development Land.*

RICS Valuation Standards, 6th edition (2007), amended (2009), 7th edition (2011) (incorporating GN2 Valuation of Individual Trade Related Property).

Robinson, J. (1989) *Property Valuation and Investment Analysis,* Sydney, Australia, Law Book Co.

Rose, J.J. (1977) *Rose's Property Valuation Tables*, Oxford, The Freeland Press.

Trott, A. (1980) *Property Valuation Methods: Interim report*, Polytechnic of the South Bank, RICS.

# Index